Biotechnology
Proteins to PCR

Cover graphic design by David Gardner, Dorchester, MA.

Biotechnology
Proteins to PCR
A Course in Strategies and Lab Techniques

David W. Burden
Donald B. Whitney

Birkhäuser
Boston · Basel · Berlin

David W. Burden
Biotechnology Training & Consulting, Inc.
P.O. Box 348
Lebanon, NJ 08833

Donald B. Whitney
Biotechnology Training & Consulting, Inc.
P.O. Box 348
Lebanon, NJ 08833

Library of Congress Cataloging-in-Publication Data

Burden, David W. (David Wilson), 1958-
    Biotechnology : proteins to PCR : a course in strategies and lab
techniques / David W. Burden, Donald B. Whitney.
        p.   cm.
    Includes bibliographical references and index.
    ISBN 0-8176-3756-7 (hardcover : alk. paper). -- ISBN 0-8176-3843-1
(pbk. : alk. paper).
    1. Proteins--Analysis--Laboratory Manuals.   2. Proteins-
-Purification--Laboratory Manuals.   3. Molecular Cloning--Laboratory
manuals.   4. Polymerase chain reaction--Laboratory Manuals.
I. Whitney, Donald B., 1955- . II. Title
    [DNLM: 1. Biotechnology--Laboratory manuals. 2. Proteins-
-analysis--experiments. 3. Proteins--isolation and purification-
-experiments. 4. Genetic Engineering--laboratory manuals.   TP
248.65.P76 B949b 1995]
QP551.B95   1995
660'.63--dc20   95-4897
DNLM/DLC   CIP

Printed on acid-free paper.
©1995 Birkhäuser Boston   *Birkhäuser* ®

ISBN 0-8176-3756-7 (hardcover)   ISBN 0-8176-3843-1 (softcover)
ISBN 3-7643-3756-7 (hardcover)   ISBN 3-7643-3843-1 (softcover)

Typeset by S. Lekowicz, Glendale, CA.
Printed and bound by Braun-Brumfield, Ann Arbor, MI.
Printed in the U.S.A.

9 8 7 6 5 4 3 2 1

To Dimples, Bristol Bay, and Johnny
D. Burden

To my parents
D. Whitney

Contents

Preface

The working environment for pharmaceutical and biotechnology scientists has seen major changes in the early 1990s. No longer are research budgets endless. The federal government, pharmaceutical industry, and academia have all tightened their belts, and downsizing has become a dreaded word in the private sector. Consequently, for those who are entering the field of biotechnology, or for those who are looking to advance their careers, the need to be valuable is never more apparent. This manual is intended to make its readers more valuable in the biotechnology laboratory.

In 1990, the Biotechnology Training Institute was established to meet the ongoing educational needs of scientists. As researchers acquired new responsibilities or wished to simply update skills, we became a source for their training. We have examined the training needs of over a thousand researchers and have used this knowledge to help develop this manual. In addition to techniques and skills, one common weakness of many of our clients is their lack of understanding of the research process, i.e., where does it start and where does it end? Similarly, from our personal experiences in teaching undergraduates, we felt that there is a lack of teaching material that covers current techniques but also gives a comprehensive view of the research process. We believe this manual addresses both needs.

We do expect students using this manual to have some background in the sciences. Obvious prerequisites should include general biology and chemistry. Course work in organic chemistry would certainly be a plus. However, microbiology and biochemistry are not necessarily prerequisites since this manual could easily accompany such lecture courses. Although this manual does provide background information and in-depth explanations on strategies and techniques, other books with a broader approach to biotechnology may be helpful references. Three good general texts are:

Glick B, Pasternak J (1994): *Molecular Biotechnology: Principles and Applications of Recombinant DNA.* Washington: ASM Press

Old RW, Primrose SB (1989): *Principles of Gene Manipulation: An Introduction to Genetic Engineering,* 4th Ed. Oxford: Blackwell Scientific Publications

Watson J, Tooze J, Kurtz D (1983): *Recombinant DNA: A Short Course.* New York: Scientific American Books

The notes and laboratory exercises in this manual evolved from several courses offered at the Biotechnology Training Institute, including Protein Purification and Characterization, Techniques of Molecular Biology, Introduction to liquid Chromatography, DNA Sequencing, and Applications of the Polymerase Chain Reaction. Tried and tested experiments from each program were assembled around a common theme. These experiments not only teach valuable skills, but also demonstrate the research process used in biotechnology laboratories.

With an understanding of academic budgets, we also wanted to develop instructional materials which were both affordable and flexible. We have endeavored to choose experiments that reflect current methodologies while minimizing cost. We have experienced the frustration of using a prepared laboratory text while not having access to the specific materials required to perform the experiments. Therefore, we have overcome this problem by focusing on a readily available organism (i.e., *Saccharomyces carlsbergensis*, or brewer's yeast) and one of its corresponding enzymes. The experiments on this organism and enzyme are not limited to the materials suggested and can be easily adapted to the desired technical level and available budget. Similarly, the subsequent cloning experiments suggest that use of particular vectors and strains, but, as indicated, alternative materials can be used to successfully perform the laboratory exercises.

We would like to thank the corporate sponsors of the Biotechnology Training Institute for providing the materials and expertise for the development of our programs, and thus for the materials in this manual. These sponsors include:

- **Barnstead/Thermolyne**, Dubuque, IA
- **Beckman Instruments**, Somerset, NJ
- **Bio-Rad Laboratories**, Hercules, CA
- **Boehringer Mannheim Corporation**, Indianapolis, IN
- **Corning Costar Corporation**, Cambridge, MA
- **FMC BioProducts**, Rockland, ME
- **Kodak Laboratory Products**, New Haven, CT
- **Labconco**, Kansas City, MO
- **MJ Research**, Cambridge, MA
- **Olympus Instruments**, Lake Success, NY
- **Pharmacia Biotech**, Piscataway, NJ
- **Savant, Inc.**, Farmingdale, NY
- **VWR Scientific**, Philadelphia, PA

We would also like to thank the following individuals for their input, comments, and suggestions: Tom Slyker, Bernie Janoson, Steven Piccoli, John Ford, Jeff Garelik, Yanan Tian, and Douglas Beecher. Special thanks to Alan Williams for his critique of the chromatography experiments and Shannon Gentile for her work in the laboratory. We would especially like to thank Maryann Burden for her comments and encouragement.

1

Introduction to the Biotechnology Laboratory

1.1 OVERVIEW

Introductory biology texts often present the biologist as a naturalist, such as Darwin or Lamark, who through careful observations develops theories and draws conclusions about living organisms. Originally these scientists kept biology and its subtopics as pure areas of study and resisted the multidisciplinary nature of modern science. For instance, in the 1830s Cagniard de Latour and von Liébig argued that fermentation was a biological phenomenon, not chemical. At this time biochemistry had yet to evolve, and the notion that biology and chemistry overlapped had not been fully realized. As the biology of the cell was discovered, several different scientific disciplines found it necessary to communicate in order to answer questions. In 1953, for instance, the structure of DNA was elucidated only after Watson and Crick pooled the information produced by biologists, chemists, and physicists. Since that time, the various scientific disciplines have continued to actively interact. However, it wasn't until the 1970s and 1980s that biology and business wholeheartedly converged to produce today's biotechnology industry.

Although biotechnology has existed for many years (e.g., the baking and brewing industries, and the microbial production of enzymes and vitamins), the explosion in interest and investment seen during the 1980s was unparalleled.

Science has become highly interdisciplinary, and, consequently, scientists require a diverse array of skills to accomplish their research. Where once a biologist might have relied on visual observation of an organism for a behavioral study, today the same research could combine visual observations with molecular techniques. It is commonplace to see a biochemist relying on recombinant proteins for analysis, a molecular biologist on a computer for data analysis, and a microbiologist on a DNA sequence for the detection of a pathogenic microorganism. Unfortunately, these disciplines are often segregated in the classroom, and true integration does not occur until graduate research. Similarly, in industry the narrow focus of research technicians will often prevent their exposure to or participation in duties outside of their immediate job responsibilities.

Our goal is to remove the barriers between scientific disciplines and to demonstrate the diverse techniques and strategies used in the biotechnology laboratory. To accomplish this, you will weave through a series of interrelated experiments designed to mimic the discovery process. The discovery process in this manual will focus on purifying and characterizing a protein and then cloning its associated gene. This manual will not only act as a source of techniques and methods involved in protein and nucleic acid research, but it also will serve as a reference and describe the research process itself.

The initial experiments presented in this first chapter will involve the common techniques of media preparation, handling and observation of yeast and bacteria, and the culturing of yeast for protein production. These exercises will lead directly to subsequent experiments on the purification and characterization of the enzyme α-galactosidase.

1.2   BACKGROUND

The process involved in creating a biotechnological product or service can take many years and involve a battery of scientists and administrators. In this, most researchers are involved in only a small portion of the discovery process. For example, in pharmaceutical research, one research team might investigate the mechanism of a disease, another could screen therapeutic natural products, a third may clone a gene encoding a valuable product, and a fourth group could be responsible for producing that product. The clinical side of this process also requires a large number of individuals and research groups.

A typical research group is staffed with junior and associate scientists and supervised by senior scientists. In many institutions, including academia, it is common for juniors and associates to be so focused in their research responsibilities that they are oblivious to other elements of the discovery process. These researchers in the trenches, however, are often an untapped resource in regard to examining and planning the research process. Administrators are now realizing that input from all participating individuals helps to avoid problems and results in a better product.

The information and experiments presented in this manual are interrelated and are presented in such a way as to be representative of an actual research project. Individuals who successfully complete the experiments in this manual will not only possess a comprehensive set of skills, but also will have repeated the equivalent of over thirty years of research previously performed by a host of individuals in numerous laboratories. Therefore this manual will serve both to illustrate the classical discovery process in biotechnology and as a source of pertinent information and techniques for individuals interested in biotechnology.

Scope of Biotechnology

Biotechnology is often thought to have arisen in the 1970s as a result of the discoveries of cloning and monoclonal antibodies. Actually biotechnology has been in use for centuries if it is viewed as any technology that employs biological systems (e.g., organisms, cells) or components (e.g., enzymes, antibodies) to achieve an applied goal. Conceptually, both making bread and medical research rely on biotechnology. It is the newsworthy (sensationalistic) and economic aspects involving genetic engineering that typically receive the greatest attention, however, biotechnology has subtly found many applications, some of which are summarized in Table 1.1.

Biotechnology is very broad based. Therefore, to be successful, a scientist will require a working knowledge of several scientific disciplines,

Table 1.1   Examples of Biotechnology

Field	Application	Comment
Agriculture	Crop improvement	Down regulation of genes encoding metabolic enzymes leads to greater shelf life of produce.
	Milk production	Recombinant bovine growth hormone (somatotropin) is fed to dairy cows to increase milk production.
	Insect resistance	Genes encoding naturally occurring insecticides can be transferred from bacteria into plants.
Environmental	Waste degradation	Microbial communities can be used to degrade sewage and industrial wastes.
Pharmaceuticals	Antibody production	Activated B-cells can be cultured in vitro for antibody production. Recombinant antibodies can also be produced by *E. coli*.
	Recombinant vaccines	Genes for selected peptides can be cloned and then used to produce large quantities of pathogen-free vaccines.
Medicine	Gene therapy	Modified DNA can be introduced into gametes and used to replace defective genes.
	Pathogen detection	Molecular probes can be used to rapidly detect species and strains of infecting organisms.
Energy	Alcohol production	Ethanol-producing microbes can be engineered to yield large quantities from low-cost substrates such as cellulose.

such as microbiology, biochemistry, immunology, etc. These skills can be combined to produce the novel products and processes often associated with biotechnology. For instance, it is impossible to design and produce a diagnostic test kit for Hepatitis B without knowledge of immunology, microbiology, and enzymology (Figure 1.1). Realistically, many scientists pool their individual expertise in the development of

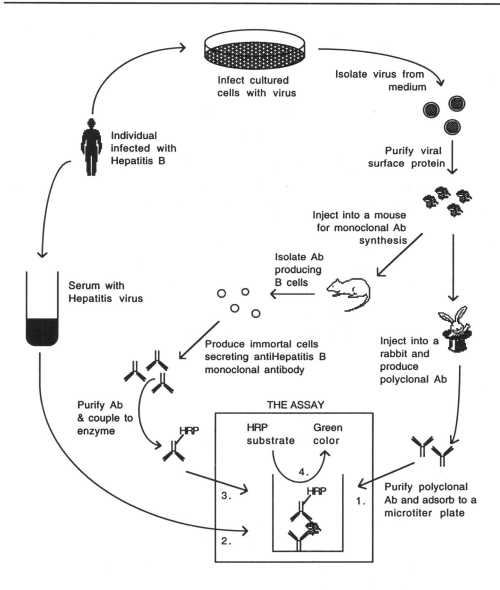

Figure 1.1 Hypothetical schematic of steps and components used in a diagnostic test kit for Hepatitis B. The steps in the assay are (1) adsorbing a polyclonal antibody to a microtiter well; (2) adding serum with Hepatitis B antigen; (3) binding an antiHepatitis B monoclonal antibody-horseradish peroxidase (HRP) enzyme conjugate; and (4) indirectly detecting the Hepatitis B by the addition of a colorimetric HRP substrate. The presence of Hepatitis B is indicated by the development of a green color.

this type of a test. This diverse collection of individuals is usually referred to as the research group.

Working in the Biotechnology Laboratory

Too often laboratory courses exclude the students from the preparation of the reagents and media needed for the experimentation. In order to make the best use of your lab time, instructors often laboriously prepare materials in advance. The students arrive during their designated lab period, deplete the supplies, create a mess, and then leave. In reality, a major portion of a researcher's time is spent setting up the experiment. Well over half of your research effort may be spent making buffers, sterilizing media, and culturing organisms. Even senior scientists can be found washing glassware. As such, these topics will be addressed to familiarize the student with some of the basic and commonplace chores associated with the biotechnology laboratory. Although commonplace, one should never underestimate the importance of establishing good laboratory practices for routine procedures.

Laboratory Safety

If you have ever been in or have seen a laboratory accident, then you know the value of lab safety. The most important aspect of lab safety is:

**USE COMMON SENSE AND PAY ATTENTION
TO WHAT YOU ARE DOING!**

Common sense notes on safety include:

- General operation: Use laboratory coats, safety glasses and gloves as appropriate. Avoid wearing tight clothing and open toe shoes.

- No Eating, No Drinking, No Smoking, No Mouth Pipetting is permitted in the laboratory!

- Pathogenic microorganisms: Nonpathogenic strains of *E. coli* and the yeast *Saccharomyces* are used throughout the program under the NIH Recombinant DNA guidelines. Though these microbes are generally regarded as safe (GRAS), please practice the following:

(1) work in a sanitary manner;

(2) employ basic aseptic technique (no mouth pipetting); and

(3) treat all waste as a potential biohazard.

▪ Hazardous chemicals such as chloroform, phenol, alcohols, and other potentially harmful reagents may be used occasionally. Note the special precautions below.

Chloroform: Organic solvent readily absorbed through the skin. Carcinogenic. Avoid vapors.

Glacial Acetic Acid: Causes severe burns. Avoid vapors.

Hydrochloric Acid: Causes severe burns. Avoid vapors.

Phenol: Toxic. Membrane and protein denaturing. Causes severe chemical burns.

Ethanol: Flammable. Poisonous.

Isopropanol: Flammable. Poisonous.

Sodium Hydroxide: Caustic. Avoid contact with the skin.

▪ Before working in the laboratory, know the locations of the eye wash, shower, safety blanket, and telephone with emergency numbers.

▪ Mutagenic chemicals: Use particular care with ethidium bromide and acridine orange as they are mutagens (use coats and gloves). Special precautions are taken for their disposal (check with your institution).

▪ Ultraviolet light: UV light can damage the retina of the eye. Particular attention should be given to the use of goggles and face shields when observing fluorescent DNA with UV irradiation. UV transilluminators can cause severe sunburn.

- Fire hazards: Note the location of fire extinguishers and blanket. Pay special attention when using flammable liquids such as ethanol.

- Disposal: Dispose of petri dishes, pipettes, pipette tips, and wastes in biological hazard bags. Special receptacles should be available for glass disposal and for syringes and needles.

- Enzymes: Though not threatening to the researcher, DNA and proteins are readily degraded by omnipresent human DNases and proteases, e.g., from the hands, sweat, and skin. The use of fresh rubber gloves is essential. Glassware should be preheated or autoclaved to inactivate stray enzymes. For special applications, glassware can be silanized to prevent DNA and protein from sticking to it. Silanization is performed by treating the glass surface with an appropriate silanizing agent, such as trimethyl-chlorosilane. Formulated preparations, such as Sigmacote™, are available commercially.

Notes, Records, and Labels

It is the duty of every laboratory instructor to impress upon the student the value of keeping an orderly laboratory notebook. Rarely, however, is the rationale for the lab notebook ever stated. Simply, the notebook is a permanent record or document of laboratory protocols and experiments. It should state how and why an experiment was performed, with sufficient detail, so that you or someone else can reproduce the results of that experiment in the future. The organization of the lab notebook should be meticulous, and include references to other protocols and information as necessary.

To illustrate the value of good notes, an analogy can be made to searching for a lost item. While traveling from your bedroom to school you unknowingly dropped your favorite ring, and the only way to find the ring is to retrace your steps. If you immediately retrace, then your short term memory will suffice. However, if you attempt to retrace your steps two years later, finding the exact path will be at best a guess. Trying to remember the protocols of an experiment is the same as retracing your steps. In the laboratory, using information recorded in a notebook limits the guesswork associated with experimentation, i.e., you limit the variables. The surest path to disaster is a sloppy notebook.

In industry, documentation is more than simply a means of minimizing guessing in the lab. Rigid protocols are often adopted so that products, especially pharmaceuticals, will receive approval for use and sale from regulatory agencies (e.g., the FDA). Researchers must adopt GLPs (good laboratory practices) while scientists in production follow GMPs (good manufacturing practices). If approved protocols are violated during the design or production of a drug, then the product may be removed from the market. Carelessness and error are not tolerated when as much as $350 million may be spent developing and testing a drug. Additionally, a complete notebook is essential for getting your research ideas patented.

For this course, as for any job, notebooks should be kept as instructed. Some institutions require specific books which are countersigned by supervisors, while others use computer-based notebooks. Records of experiments, such as photographs, printouts, chromatograms, etc., should be taped directly into notebooks. Pertinent information (date, experiment, trial) should be written directly on the record. Scanners can be used to incorporate records directly into computers. Never copy data on a paper towel and incorporate it into the notebook later. (This is all too common.)

Keeping bottles and tubes accurately labelled can be just as important as an organized notebook. Most people working in a laboratory, at one time or another, have placed an unlabelled tube in the freezer or refrigerator, knowing they would return to it shortly. When that tube reemerges several weeks, months, or years later, the contents are unknown (it could be water or botulism toxin). Tubes, organisms, media, and reagents must be adequately labelled to avoid such confusion. Not only is it important to keep tabs on what is what, but federal and state laws can mandate that all containers in a laboratory be adequately labelled.

Housekeeping

Many aspects of experimentation are highly dependent upon laboratory maintenance. Not maintenance in the sense of trash disposal and floor mopping, but in regard to glassware preparation, cleaning of work areas, handling of chemicals, storage of materials, and care of equipment.

Washing Glassware

A standard protocol used to wash glassware includes: (1) wash the item with a laboratory grade detergent and brush; (2) rinse three times with tap water; (3) rinse three times with distilled or deionized water; and (4) invert and air or oven dry. If the glassware requires sterilization, dry heat ovens or an autoclave can be used. Ovens have the advantage that all moisture is removed thus preventing microbial growth in the condensation often found in autoclaved materials. However, dry heat does dry and crack rubber seals and plastic caps whereas an autoclave will not.

Decontamination

Surfaces exposed to microbes should be washed with a sanitizing agent (e.g., commercial solution or 70% ethanol). When possible, disposable lab matting or mattress pads can be used to cover a bench. After use, these pads can be autoclaved and discarded. On general surfaces, molds will grow in humid areas, and spores are present in all dust. Thus, damp areas should be disinfected regularly, and dust should be removed.

Handling and Storage of Chemicals

It is important to establish a clear protocol for the handling and storage of stock chemicals, (i.e., purchased chemicals). Since most stock chemicals are dry salts, they can be stored at room temperature. Many reagents, such as acids, bases, and solvents should be stored in vented storage cabinets. Flammable solvents must be stored in explosion proof storage cabinets. When using these stocks, only clean spatulas and pipettes should enter the containers, and any unused reagent taken out of its container should not be replaced (due to the risk of contamination). It is a good idea to separate these stocks from the solutions they are used to formulate.

Many solutions can be prepared in advance of an experiment and stored. Inert solutions, such as 5 M NaCl, will not spoil and can remain indefinitely on a shelf as long as it is sealed well. However, other solutions will go bad, especially those that will support microbial growth such as glycine and citrate. Take the time to differentiate between stable solutions and those that will degrade or support microbial growth. Often, autoclaving a solution before storage will prevent its spoiling.

Equipment Maintenance

Nothing is more frustrating than working in a lab filled with disabled equipment. When instruments have multiple users, it is common for equipment to be abused and break. Maintaining functional equipment can only be accomplished if individuals take responsibility for instrumentation they use and if supervisors ensure that all users are adequately trained. When equipment does break (which even well maintained equipment will do), then a mechanism should be in place for its rapid repair. Inoperable equipment results in the loss of productivity and money.

Preparation and Storage of Buffers

Biological materials require buffered solutions. Consequently the preparation and storage of buffers and other solutions (salt, organic) is fundamentally important. There are several conventions used to prepare buffers, for instance:

- Molar Solutions (M)—number of moles of a substance in a total of 1 liter of solvent,

- % Weight/Volume (w/v)—number of grams of a substance per 100 ml of solvent,

- % Volume/Volume (v/v)—number of milliliters of a solution in a combined total of 100 ml of solvent.

It is important to realize that researchers will define these conventions many different ways. Furthermore, to confuse the issue, more than one of these measurements can be found in a single buffer. For example, the solution 50 mM Tris, pH 7.5, 0.1% SDS, 0.05% Tween 20, contains Tris (M), SDS (w/v), and Tween 20 (v/v). The ability to decipher the components of a solution are gained with experience.

Many buffers are inert, but some may degrade or support microbial growth. If a buffer is labile, then it should be made as needed. However, solutions which can be prepared in bulk (and as concentrates) are often convenient. Certain buffers, such as those containing the amino acid glycine or the carboxylic acid citrate, are wonderful sources of energy for microbial contaminants and should be sterilized prior to storage.

Preparation and Storage of Media

Media is the plural of medium, the substance used to cultivate organisms. Different organisms require different media, ranging from a simple salts and glucose solution for bacteria to the complex sera containing media for animal cell culture (Table 1.2).

The means by which a medium is prepared are completely dependent upon its components. For media containing simple sugars and salts, dissolving the components in water followed by autoclaving in a loosely capped container is satisfactory. However, more complex solutions may contain heat labile components, such as antibiotics and growth factors, and consequently can not be autoclaved. This media can be filter sterilized or irradiated. One trick is to mix and autoclave the heat stable components first and then add filter sterilized labile components after the autoclaved solution has cooled. When concocting media for the first time, it is best to follow the recipes found in research articles and manuals.

Table 1.2 Some Organisms and Their Media

Organism	Medium	Comments
Escherichia coli	LB broth	Simple media made of tryptone, sodium chloride, and yeast extract.
Saccharomyces cerevisiae	YPD broth	Medium containing yeast extract, peptone, and glucose.
L929 mouse cell line	Dulbecco's Modified Eagle's Medium (DMEM) with 10% fetal calf serum	A relatively complex synthetic medium which is supplemented with sterile fetal calf serum. DMEM contains glucose, vitamins, and minerals while the fetal calf serum provides hormones and growth factors needed by mammalian cells

Care and Maintenance of Cultures

Microorganisms that are continuously used for research need to be maintained, i.e., kept fresh. Maintenance can be accomplished by subculturing organisms onto fresh medium on a routine basis (i.e., once or twice a week).

The techniques used for culturing are dependent upon the organism. Bacteria and yeast are easily cultured by streak plating, while animal cell culture requires subculturing, which is the dissociation of cells from the culture flask with trypsin, dilution of the cells with a serum based medium, and transfer of cells to new flasks. Due to the complex and expensive nature of cell culture, this manual will not include these techniques.

The most common method of maintaining microbes is to streak them onto solidified agar media. Prior to working with any microorganism, the cultures must be grown and pure. Streaking microbes onto agar plates is one easy method of separating individual cells and allowing them to develop into isolated colonies (Figure 1.2).

Between each streak, the loop is sterilized with a flame, allowed to cool (it may be touched to a sterile area of the agar), and then used to spread or dilute the microorganisms over the surface of the plate. Streaked plates are inverted during incubation so that condensation that forms on the lid doesn't drip onto the surface of the plate. Throughout this procedure, aseptic technique is used.

Aseptic technique is used to prevent contaminating cultures and sterile items. Though it sounds like a protocol, aseptic technique is more conceptual (a frame of mind) than otherwise. It is important to realize that all equipment, solutions, gases, and other items (especially hands) are laden with microorganisms and enzymes that are potential contaminants. Aseptic technique recognizes these risks and avoids exposing pure cultures or sterile systems to items which are not sterile. Some simple guidelines for aseptic technique include:

■ Never needlessly expose a sterile object to a nonsterile object, especially air.

■ All work surfaces should be cleaned and sanitized. Never place caps from culture flasks or media bottles onto a surface. Containers should be opened and closed quickly.

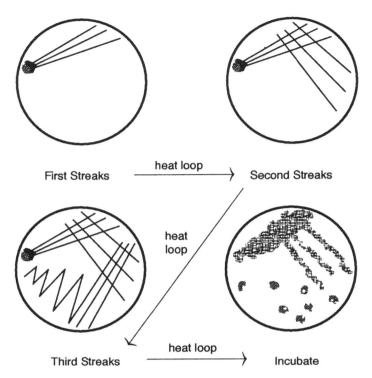

Figure 1.2 Pattern followed when streaking an agar plate with an inoculation loop.

- All solutions used to culture cells must be sterile. Sterilization can be accomplished by filtration, autoclaving, or irradiation.

- All cultureware, e.g., pipettes, flasks, dishes, must be sterile. It is advisable to use disposable materials when available.

- Gloves should always be worn to protect the cultures from microbes present on hands.

- Don't breathe on open culture flasks and dishes!

The number of organisms used in recombinant DNA research is continuously increasing, with species ranging from phages to yeast and geraniums to mice. *Escherichia coli*, the molecular biologist's workhorse, itself has numerous variants and strains. Each variant possesses a genotype that makes it valuable for particular line of experimentation (Table 1.3).

Deciphering the genetic nomenclature used to describe organisms can be complicated. Organisms that are regularly manipulated, e.g., *E. coli* and *S. cerevisiae*, have been extensively mutated for both biological control and experimental purposes.

When discussing the genetic characteristics of these organisms, it is important to differentiate between phenotype and genotype. Phenotype is the visible characteristics of a gene, e.g., resulting color, enzyme activity, growth characteristics, sugar assimilation or fermentation, etc. Genotype refers to the actual genetic make-up of an organism, and such characteristics as dominant, recessive, wild type, and mutant. The symbols and terminology used to describe phenotypes and genotypes of *E. coli* appear in Table 1.4.

Handling and storing microbial strains correctly is extremely important in molecular biology (as it generally is in microbiology). *E. coli* strains are genetically modified to be sickly (a form of biological control) and tend to die easily. Even *E. coli* streaked on LB agar and stored at 4°C can die within several weeks. Yeasts, though generally more hearty, are also much the same. Their care includes the following techniques.

Table 1.3 Several Commonly Used *E. coli* Strains and Their Characteristics

Strain	Characteristics	Application
HB101	F$^-$, recA13, hsdS20($r_B^-m_B^-$), supE44, λ^-, leuB6, ara-14, proA2, lacY1, galK2, rpsL20(str^r), xyl-5, mtl-1	General cloning vehicle
DH5α	F$^-$, recA1, hsdR17($r_k^-m_k^+$), supE44, thi-1, λ^-, gyrA96, Δ(lacZYA-argF)U169, φ80dlacZΔM15, end A1	General cloning with α complementation (blue/white selection)
JM109	F', traD36, proAB$^+$, LacI$_q$ZΔM15, Δ(Lac-proAB), supE44, hsdR17($r_k^-m_k^+$), gyrA96, recA1, thi, endA1, relA1, λ^-	M13 vector host for single-stranded DNA synthesis

Table 1.4   Key for *E. coli* Genotypes and Phenotypes

Genotype	Phenotypic Characteristic
ara-14	Mutation that prevents arabinose metabolism.
Δ(lacZYA-argF)U169	A deletion of DNA consisting of the *lac* operon through the *argF* gene.
F⁻	Lacks the F plasmid which is responsible for pilus formation and conjugation.
lacZΔM15	Deletion of the lacZ alpha peptide. This deletion results in an inactive β-galactosidase protein due to the absence of an alpha peptide.
galK2	Lacks the ability to utilize galactose.
gyrA96	Mutation in a subunit of DNA gyrase which results in the resistance to the antibiotic nalidixic acid.
hsdR17($r_k^-m_k^+$)	Functional mutation in the DNA cleaving subunit of the EcoK restriction enzyme.
hsdS20($r_B^-m_B^-$)	Mutation in the DNA sequence recognition subunit of the EcoK restriction modification system. This mutation prevents both DNA cleavage and methylation.
λ⁻	Lacks a lysogenic (latent and integrated) Lambda bacteriophage.
lacIq	Mutation that results in the overproduction of the repressor protein for the *lac* operon.
leuB6	Auxotrophic for leucine biosynthesis.
mtl-1	Deficient in the metabolism of mannitol.
proA2	Auxotrophic for proline.
recA1, recA13	Deficient for the enzyme responsible for strand exchange in recombination.
rpsL20(str^R)	Mutation in a ribosomal protein that results in streptomycin resistance.
supE44	Substitutes a glutamine for an amber mutation, i.e., a mutation prematurely halting translation.
thi-1	Thiamine auxotrophy.
traD36	Mutation in the gene responsible for transferring DNA through a sex pilus.
xyl-5	Inability to metabolize xylose.

Stabs and Slants

Cultures may be stored in tubes or vials containing agar, although it is not recommended for long periods of time for *E. coli*. The method simply involves streaking the agar surface (usually in a sterile screw capped vial) with an inoculation needle and then stabbing the needle into the agar. After allowing the culture to incubate overnight, the vials are screwed shut and sealed with Parafilm to prevent the agar from drying out. Agar slants in test tubes are often used to preserve yeasts in a similar manner by simply streaking yeasts onto the surface of the slant. After the culture grows, sterile mineral oil can be used to cover the slant to prevent it from drying. However, the bacteria and yeast strains used in recombinant DNA research are usually weak and do not lend themselves to these storage methods of classical microbiologists.

Freezing

Most cultures are easily preserved frozen, either in liquid nitrogen or in a –80°C freezer. However, the freezing of microorganisms tends to cause cell lysis due to the intracellular formation of ice crystals. The addition of glycerol to 10% (v/v) prior to freezing cells helps to prevent cell lysis and thus preserves the cultures.

A simple method of preserving both yeast and *E. coli* involves preparing sterile cryogenic vials (e.g., Corning Costar 2.0 ml screw capped tubes) with small glass beads (Fisher Scientific 2 mm glass beads) and a 20% solution of glycerol (i.e., ⅓ full of beads and with 250 μl of 20% glycerol). When cells are cultured, strains can be maintained by simply adding 250 μl of culture broth to a prepared vial, vortexing briefly, labelling, and freezing at –80°C. **The importance of establishing a simple practice such as this for maintaining commonly used cultures cannot be emphasized strongly enough.** To reactivate a culture, simply remove the vial from the freezer (keep on ice for a minimum of time), and with a sterile inoculating needle, break away several glass beads, pour the freed beads onto an agar plate, roll the beads around on the agar surface, and incubate the plate. A smear of cells usually develops which provides an inoculum for subsequent streaking.

Lyophilization

Many organisms can be stored by freeze drying, a process that involves freezing the cells (or other substance) and then removing moisture with a vacuum pump. This process allows water in the cells to transfer between a solid phase and gaseous phase directly, i.e., sublimation. Not all organisms can be lyophylized, e.g., cultured animal cells and fragile microbes (including some strains of *E. coli*), but many standard organisms are readily preserved in this manner.

General Rules for Protein Handling

Proteins are large, complex biomolecules that are extremely sensitive to denaturation (i.e., loss of native conformation and, consequently, loss of biological activity) under typical laboratory conditions. The loss of biological activity can be partial or complete depending on the extent of the denaturation. Manipulation of proteins in the laboratory requires the researcher to follow special rules to avoid contaminating and denaturing samples. The following rules also apply to handling enzymes.

The first rule of protein handling is to wear gloves. Proteases from the skin can degrade the sample as well as contaminate an otherwise pure protein solution.

Mixing and dissolving proteins should be done under mild conditions. Generally, proteins solutions should not be vigorously stirred because the shear forces that would be generated can degrade the integrity of the protein molecule. Additionally, protein solutions tend to foam if stirred or vortexed too vigorously.

Generally, protein solutions should be maintained at high concentration (ideally > 1 mg/ml) as the protein is usually more stable. Proteins tend to stick to most surfaces, and this can be a serious source of protein loss during manipulation and handling, especially with dilute solutions. Loss of protein in this way can be minimized by maintaining a high concentration or by adding a second inert protein, such as albumin, which increases the overall protein concentration. The inert protein should be one that can be easily removed later on.

Glassware or plasticware used in protein studies should be meticulously washed and rinsed with deionized water. Frequently, treatment with EDTA before the water rinse will lower the chances of contamination of the protein by metal ions. Silanization of glassware helps to prevent protein adsorption to the glass.

Long and short term storage of proteins present many potential stability problems. Proteins can generally be stored either frozen, at 4°C, or lyophilized. Ultralow temperatures, e.g., −80°C, can irreversibly denature some proteins. When working with protein solutions in the laboratory for any appreciable length of time, the solutions should be kept cold. This can be done by keeping tubes and bottles on ice.

Proteins can be stored short term in a typical 4°C refrigerator providing that the buffer conditions, i.e., salt concentration, pH, reducing environment, etc., support protein stability. Sometimes addition of a protease inhibitor is required to minimize degradation. Addition of stabilizers such as glycerol may also be necessary to maintain protein activity (maintenance of activity is usually a good measure of protein stability). The presence of glycerol (typically < 20%) generally promotes protein stability and will prevent the protein from freezing, even at temperatures below 0°C. Addition of glycerol, 50% v/v, is useful for long term storage but may present problems in any subsequent purification procedures (e.g., chromatography). The use of antibacterial agents, such as sodium azide, is appropriate in order to limit microbial growth during storage at higher temperatures.

Although many proteins are quite stable when stored frozen, repeated freezing and thawing can result in loss of activity. During the freezing process the protein may be subjected to extremes of pH and/or buffer salt concentrations which may result in denaturation. This effect can be minimized by rapidly freezing the protein solution, usually by submersion in a dry ice bath of either ethanol or acetone. The frozen protein solution can then be safely stored at −20°C. Storage of proteins by freezing should be considered a long term storage solution and repeated freezing/thawing should be avoided. One approach to avoid freeze/thawing is to divide the protein sample into aliquots. In this way, each aliquot has to be thawed only once, thus preventing serious loss of activity. This technique can be used for other nonprotein, unstable molecules, such as ammonium persulfate solutions.

Lyophilization is a very effective method of protein storage. The protein solution is frozen (rapidly!) and then placed in a lyophilizer. Under a vacuum, the frozen liquid sublimes leaving the protein behind, usually as a fluffy white solid. The protein to be lyophilized must be dissolved in either water or a buffer solution containing volatile buffer salts that also sublime under the lyophilization conditions. The drawback to lyophilization is the occasional difficulty in redissolving the protein or in recovering full activity.

Storage of Nucleic Acids

DNA is generally a very stable molecule and can be stored for years either desiccated (dry) or in solution (frozen). Environmental nucleases can destroy DNA rapidly, however, and care needs to be exercised not to contaminate solutions. These nucleases require Mg^{+2} for activity; thus dissolving DNA in solutions containing EDTA chelates the Mg^{+2} and helps to preserve the sample. DNA that is contaminated with nucleases either disappears from solution, or, when assessed on an agarose gel, appears smeared or with an altered electrophoretic pattern (as compared to the original pattern). Autoclaving solutions will destroy most, but not all nucleases. DNA solutions can be dried in a SpeedVac® which remove liquid and yield crystalline DNA. Tubes can be capped and stored at room temperature, or more preferably at 4°C or –20°C.

Due to extremely resilient RNases, it is more difficult to isolate and store RNA than DNA. RNases tend to be thermostable and are more difficult to destroy. These enzymes are present in cells and ubiquitously on glassware, instruments, in water, and in solutions. Several precautions can be followed to prevent ribonuclease contamination, such as baking glassware, treating solutions with diethyl pyrocarbonate (DEPC), and working in conditions that are free of ribonuclease contaminants. (Caution: DEPC is a carcinogen!) RNases arising from cells can be controlled by adding ribonuclease inhibitors, such as vanadyl complexes or RNasin. Individuals working with RNA are encouraged to review Blumberg (1987).

1.3   EXPERIMENTAL DESIGN AND PROCEDURES

The experimental side of this manual will focus on the purification of α-galactosidase from *Saccharomyces carlsbergensis* and the cloning of its

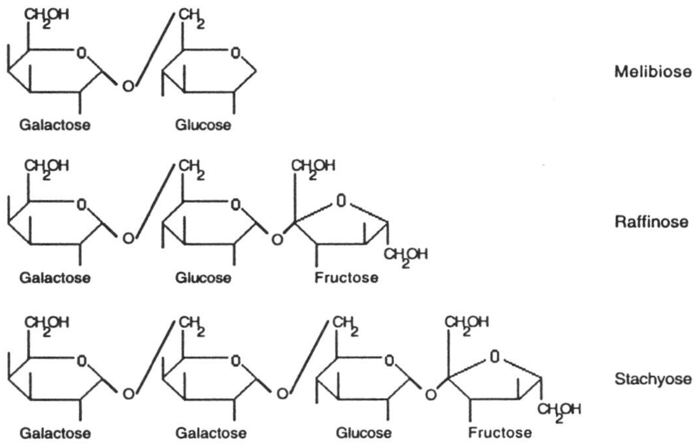

Galactose   Glucose   Melibiose

Galactose   Glucose   Fructose   Raffinose

Galactose   Galactose   Glucose   Fructose   Stachyose

Figure 1.3 Structures of melibiose, raffinose, and stachyose.

associated gene (*MEL1*). α-Galactosidase is an enzyme that cleaves the alpha anomeric linkage between galactose and an adjacent moiety. Melibiose, raffinose, and stachyose are all α-galactosides (Figure 1.3).

α-Galactosidase is a glycoprotein localized outside the cell membrane. The native enzyme is a dimer with a molecular weight of 270,000 daltons and is approximately 57% carbohydrate. The enzyme has been isolated and purified by Lazo et al (1977, 1978) using a combination of anion exchange and gel filtration chromatographies. α-Galactosidase is a ubiquitous enzyme which has been found in bacteria, molds, yeast, plants, and animals. In humans a deficiency of α-galactosidase is responsible for the onset of Fabry disease which is a fatal metabolic disorder.

α-Galactosidase has many commercial applications. The enzyme is valuable in the processing of sucrose from sugar beets due to the ability of α-galactosidase to hydrolyze raffinose which in beet juice diminishes the rate of sucrose crystallization. Baker's yeast produced on beet molasses can be supplemented with α-galactosidase in order to increase carbohydrate availability and processing efficiency. The α-galactosidases from *Aspergillus* spp. are available for human consumption. The enzyme is mixed with food and facilitates the removal of stachyose and other flatulence causing α-galactosides prior to reaching the intestine. Recombinantly produced human α-galactosidase has application for the treatment of Fabry disease.

Yeast α-galactosidase and the *MEL1* gene are going to be the vehicles used to demonstrate protein purification and characterization as well as gene cloning. In this initial chapter, the experiments are aimed at familiarizing you with the laboratory and preparing for future experiments. The techniques covered in this section will be used throughout the manual; therefore, take the time to thoroughly learn these basic skills. It is important to have confidence in the materials you are preparing so that you can have confidence in the results of the subsequent experiments.

Preparation of YPD and YPG Media

You will prepare two different liquid media for the cultivation of yeasts, namely YPD and YPG (i.e., yeast (extract), peptone, dextrose, and yeast (extract), peptone, galactose). The media will be used for the initial cultivation and analysis of α-galactosidase synthesis from *S. carlsbergensis*.

Materials

Graduate cylinder (50 ml)
Beaker (100 ml)
Two flasks (125 ml)
Foam or cotton stoppers
Aluminum foil
Autoclave tape
Yeast extract
Peptone
Glucose
Galactose
Deionized water

Method

1. Glassware and stoppers should be detergent free. Wash these as previously described if necessary.

2. General media preparation simply involves weighing a percentage of the volume (in grams) of each reagent and then adding that component to the water. Unlike molar solutions, the change in volume

due to the solutes is usually considered unimportant. YPD is made by dissolving yeast extract (1%), peptone (2%), and glucose (2%) into deionized water. These components should easily dissolve with stirring. For this experiment, you will need 25 ml of YPD; therefore mix 0.25 g yeast extract, 0.5 g peptone, and 0.5 g glucose into 25 ml of deionized water. If this experiment is to be performed by a group of twenty-five or less, one batch will easily be sufficient for the entire group.

3. Transfer the medium to a flask.* Plug the flask with the foam or cotton stopper. These stoppers allow for gas exchange but prevent microbial contamination. Cover the stopper and neck of the flask with foil. Place a small piece of autoclave tape on the foil. The foil prevents dust from accumulating on the opening and neck of the flask following autoclaving.

4. Repeat the above formulation substituting galactose for glucose.

5. Place the flasks in an autoclave and sterilize by heating to 121°C for 15 min. *If you are unfamiliar with the autoclave, find someone to instruct you on its proper use.* After the autoclave cools, remove the flasks and allow them to cool to room temperature. The autoclave tape should indicate the flasks reached the proper temperature.

6. The media can be stored at room temperature until use.

Option: In addition to the general sugar, peptone, yeast extract based media, defined media can also be assessed for α-galactosidase production. One common defined medium is based on Yeast Nitrogen Base (from Difco) which contains all essential nutrients for wild type yeasts. Yeast Nitrogen Base contains the following ingredients per liter:

---

*These flasks will eventually be inoculated with yeast and cultivated aerobically. This is normally accomplished by placing the culture in an incubated shaker (or shaker in an incubator room). If a shaker is not available, a stir bar can be placed in the flask prior to autoclaving. After the culture is inoculated, the flasks can be placed on magnetic stir plates either at room temperature or in an incubator. The stir bar will keep the culture adequately aerated to ensure good growth of the yeast.

Ammonium Sulfate	5 g	l-Histidine HCL	10 mg
dl-Methionine	20 mg	dl-Tryptophan	20 mg
Biotin	2 µg	Ca Pantothenate	400 µg
Folic Acid	2 µg	Inositol	2 mg
Niacin	400 µg	p-Aminobenzoic Acid	200 µg
Pyridoxine HCl	400 µg	Riboflavin	200 µg
Thiamine HCl	400 µg	Boric Acid	500 µg
Copper Sulfate	40 µg	Potassium Iodide	100 µg
Ferric Cl	200 µg	Manganese Sulfate	400 µg
Sodium Molybdate	200 µg	Zinc Sulfate	400 µg
KH_2PO_4	1 g	Magnesium Sulfate	0.5 g
NaCl	0.1 g	$CaCl_2$	0.1 g

Yeast Nitrogen Base (YNB) is simply combined in solution with a desired sugar and then sterilized as described above. However, due to the minimal nutritional nature of this type of media, yeast tend to grow more slowly as compared to a rich yeast extract, peptone based media. Yeast cultured on defined media generally require 4 to 5 days incubation to yield comparably dense cultures. One major limitation of defined media is that many genetically modified yeasts (e.g., yeasts with mutations in uracil, leucine, adenine, and tryptophan biosynthetic pathways) are incapable of growth without nutritional supplements. For 25 ml of YNB glucose medium, dissolve 0.27 g Yeast Nitrogen Base and 0.25 g of glucose in water. Sterilize as described above. YNB galactose medium is prepared accordingly.

Streak Plating of *Escherichia coli* and *Saccharomyces carlsbergensis*

Streaking microorganisms is routine in the biotechnology laboratory. Streak plating employs aseptic technique. *Do not touch or needlessly expose any sterile item to a nonsterile item, especially AIR.*

Materials

Gas burner or alcohol lamp

Inoculation loop

MORE...

LB agar plate—If this is not available, it can be made by dissolving 1% tryptone, 1% NaCl, 0.5% yeast extract, and 2% agar in water. Sterilize by autoclaving and aseptically pour into disposable plastic petri dishes. Five plates can easily be poured from 100 ml of liquified LB agar.

YPD agar plate—YPD agar plates are 2% glucose, 2% peptone, 1% yeast extract, and 2% agar. Prepare similarly to LB agar plates.

Escherichia coli stock culture—For comparative purposes, any strain will be suitable.

Saccharomyces carlsbergensis stock culture—This yeast will be used for the production of α-galactosidase and as a donor of DNA for cloning. *S. carlsbergensis* is synonymous with *S. uvarum*, or more commonly, brewer's yeast. A hearty, wild type strain with good growth characteristics should be used. Strains can be obtained from the American Type Culture Collection, or less expensively, from commercial hobby shops specializing in home brewing.

Method

1. Arrange yourself with the cultures and flame within close reach. Place the LB agar plate facedown in front of you.

2. Taking the inoculating loop in one hand and the *E. coli* culture tube in the other, heat the inoculating loop, remove the culture tube plug (cap) with the backside of your free fingers, quickly and briefly flame the opening of the culture tube, insert the loop and remove a small amount of cells (or broth), flame the opening, and replace the plug (cap). The culture tube should not be open for more than 5 seconds. You should also hold your breath any time a culture tube or petri dish is open.

3. Grab the bottom of the LB agar plate and lift it up exposing the agar. Smear the *E. coli* in a small spot on the agar surface. Close the plate. The pressure you exert on the agar should be minimal since agar will tear easily.

4. Flame the inoculation loop, open the plate, and streak the *E. coli* as illustrated in Figure 1.2. Flame the loop between *each set* of streaks. Also flame the loop after the final streak so to kill any residual bacteria.

5. Keeping the plate inverted, place it in a 37°C incubator. Inverting the plate prevents condensation from forming on the lid and dripping onto the surface of the plate. The *E. coli* can be incubated at room temperature if necessary.

6. Repeat the streaking with the yeast. Yeast are typically incubated at 30°C. They may also be grown at room temperature if necessary.

7. Cleanup your work area with a suitable disinfectant.

Microscopic Observation of Bacteria and Yeast

It is very important to periodically check your organisms to be sure that they are what they are supposed to be. It is not uncommon for molecular biologists to work with *E. coli* and not have access to a microscope. Cultures can become contaminated and thus require monitoring. Simple microscopic observation and cell staining are useful tools for maintaining culture integrity.

Materials

Yeast and *E. coli* cultures (streak plates)
Slides
Cover slips
Alcohol lamp or gas burner
Inoculation loop
Water
Microscope with at least 40× objective and 10× ocular lenses
Optional: Gram stain kit
Optional: Gram positive organism, such as *Bacillus subtilis*

Method

1. Place a small drop of water (10–20 µl) on a glass slide. Using aseptic technique, transfer an inoculation loop of *E. coli* from the culture tube to the slide. Cover the drop with a cover slip.

2. Examine the microbes at 100×, 400×, and 1000×. Since the bacteria are not stained, they may be difficult to see. If so, reducing the light intensity by closing the lamp condenser helps to create contrast. Your objective is to develop a conceptual idea of the size and shape of the *E. coli*. Temporarily save this slide.

3. Repeat the above steps and examine the yeast. Yeast are much larger and more easily seen. Return to the first slide for comparison.

4. Option: Perform a Gram stain on the *E. coli*. The Gram stain is a major technique used to differentiate between the two major groups of bacteria, i.e., Gram positive and Gram negative. It can also be used to monitor your *E. coli* cultures for contamination. The staining protocol is as follows:

   a. Smear a diluted sample (10 µl) of *E. coli* onto a slide and allow to air dry.

   b. Quickly pass the slide through a flame (Bunsen burner or alcohol lamp) four or five times. The heat fixes the cells to the slide.

   c. Cover the cells with Crystal Violet solution. Incubate for 30 sec. Pour off the stain and rinse with water (use a wash bottle).

   d. Cover the cells with iodine solution and incubate for 60 sec. Wash gently with water and drain.

   e. Cover the cells with 95% ethyl alcohol and incubate for 15 sec. Pour off and gently rinse with water.

   f. Cover the cells with the counter stain Safranine. Incubate for 30 sec and again rinse with water. Dry the slide on a paper towel or with blotting paper (don't wipe off the cells).

   g. Examine the cells at 400× to 1000× under a microscope. *E. coli* are Gram negative and appear as pink cells. Gram positive cells are purple.

5. Option: If a Gram positive bacteria is available, such as *Bacillus subtilis*, Gram stain and compare to the *E. coli*.

Inoculation of YPD and YPG with *Saccharomyces carlsbergensis*

The synthesis of α-galactosidase by yeast is dependent upon the sugars in the culture media. *S. carlsbergensis* is to be inoculated into YPD and YPG media (prepared earlier), cultured, and then screened in the subsequent chapter for the presence of α-galactosidase.

Materials

Sterile YPD medium (25 ml in a 125 ml flask)
Sterile YPG medium (25 ml in a 125 ml flask)
Streak plate of *S. carlsbergensis*
Inoculating loop
Bunsen burner or alcohol lamp
Incubator shaker or equivalent

Method

1. To ensure that the yeast will be grown as pure cultures, use aseptic technique. Wash the lab bench with a disinfectant. Position the media, burner, streak plate, and loop conveniently on the bench. Light the burner.

2. Heat the loop until it glows. The front half of the handle may be sanitized by briefly passing it through the flame.

3. Open the petri dish (briefly!) and remove a small portion of an isolated yeast colony. If the needle is still hot from the flame it can first be cooled by touching it to a sterile area of the agar.

4. Remove the foil from the YPD flask, remove the plug (grasp it in the center so as not to contaminate the lip of the flask), and insert the tip of the loop (with the cells) into the broth. **Do not touch the sides of the flask with the loop as you cannot guarantee it is sterile.** Remove the loop and plug the flask. During the inoculation, the plug should

never be placed on the bench. This inoculation should take less than 5 seconds.

5. Heat the loop to kill any residual cells.

6. Inoculate the YPG flask in the same manner.

7. Place the flasks on an incubator shaker at 30°C for 48 hr. If necessary, the cultures can be stored at 4°C for several days before analysis.

STUDY QUESTIONS

1. What precautions could you take to ensure the preservation of an important microbial culture?

2. If a culture of *E. coli* HB101 (Table 1.3) was thought to be contaminated by *E. coli* DH5α, what could you do to resolve this question?

3. Investigate the metabolic pathway and genetics of galactose utilization in yeast.

FURTHER READINGS

Phaff HJ, Miller MW, Mrak EM (1978): *The Life of Yeasts*. Cambridge, MA: Harvard University Press

Atlas R (1988): *Microbiology: Fundamentals and Applications*. New York: Macmillan

REFERENCES

Blumberg D (1987): Creating a ribonuclease-free environment. *Methods in Enzymology* 152:20–24

Lazo PS, Ochoa AG, Jascón S (1977): α-Galactosidase from *Saccharomyces carlsbergensis*: Cellular localization, and purification of the external enzyme. *Eur J Biochem* 77: 375–382

Lazo PS, Ochoa AG, Jascón S (1978): α-Galactosidase (melibiase) from *Saccharomyces carlsbergensis*: Structural and kinetic properties. *Arch Biochem Biophys* 191: 316–324

2

Introduction
to Proteins

2.1 OVERVIEW

The obvious first step in biochemical research is the discovery of some
interesting biological activity followed by efforts to determine what type
of biomolecule is responsible for that activity. The next few chapters in
this book will deal with some of the issues that arise when the molecule
of interest is a protein that must be purified and characterized prior to
cloning the associated gene.

The structure and composition of a protein are fundamental to its
biochemical characteristics which are exploited for its purification. The
discussion of protein biochemistry in the background section of this
chapter is presented as a review of the fundamentals and is not intended
to be comprehensive.

The structure, function, and synthesis of the enzyme α-galactosidase
is the model used for the protein purification exercises presented in this
manual. Though this purification is of an enzyme, the protocols and techniques used can be applied to most proteins, regardless of function. The
generic term protein is often used in the background sections of this

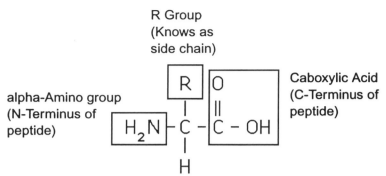

Figure 2.1 Structure of an α amino acid.

manual while the more specific term enzyme is used for the experimental sections.

Previously, *S. carlsbergensis* has been cultured and inoculated into both YPD and YPG media. The synthesis of α-galactosidase by *S. carlsbergensis* will be assessed in order to determine the most appropriate medium for the production of the enzyme. The biological activity (i.e., hydrolysis of galactose-containing sugars such as melibiose and tachyose) will be measured by hydrolysis of the substrate, p-nitrophenyl-α-D-galactoside. Additionally, the proteins present in the yeast broth will be measured by several colorimetric based assays to determine total protein concentration.

2.2 BACKGROUND

Proteins are large, polymeric biomolecules that are composed of α amino acids covalently linked through peptide bonds. An α amino acid (Figure 2.1) has both acid ($-COO^-$) and amino ($-NH_3^+$) groups bonded to the same carbon atom, while the peptide bond is the condensation product between two α amino acids (Figure 2.2).

Although each species of protein possesses the same amino acid sequence, proteins are considered heterogeneous polymers because the amino acids which comprise the sequence of the protein are not identical. Proteins are typically composed of some combination of the twenty naturally occurring amino acids (Table 2.1).

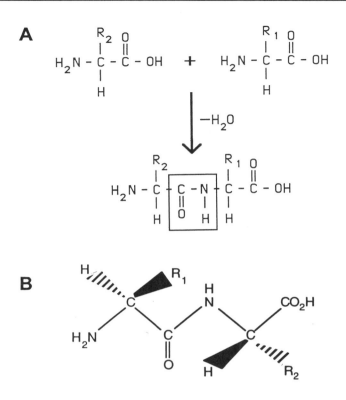

Figure 2.2 (A) Peptide bond formation. (B) Three dimensional view of dipeptide showing planar peptide bond.

All biomolecules are essential in their own right, but proteins take on special importance due to the versatility of their biological functions. Proteins regulate metabolic activity, catalyze the biochemical reactions necessary for proper cell function, and maintain the structural integrity of the cell and organism. These important biomolecules can be classified by, among other things, their chemical composition, structure/conformation, and biological function.

Based on chemical composition, proteins can be divided into simple proteins and conjugated proteins. The so-called simple proteins are composed solely of some combination of the twenty naturally occurring

Table 2.1   The Twenty Naturally Occurring α Amino Acids Commonly Found in Proteins

Amino Acid	Abbrev	Side Chain	Type	pK$_a$	Mol Wt
general		(general structure)		carboxy 3.1 amino 8.0	
alanine	ala, A	$-CH_3$	hydrophobic		89.09
arginine	arg, R	$-CH_2CH_2CH_2NHC(NH)NH_2$	basic	12.48	174.2
asparagine	asn, N	$-CH_2CONH_2$	hydrophilic		132.1
aspartic acid	asp, D	$-CH_2COOH$	acidic	3.86	133.1
cysteine	cys, C	$-CH_2-SH$	hydrophilic	8.33	121.12
glutamic acid	glu, E	$-CH_2CH_2COOH$	acidic	4.25	147.13
glutamine	gln, Q	$-CH_2CH_2CONH_2$	hydrophilic		146.15
glycine	gly, G	$-H$	hydrophilic		75.07
histidine	his, H		basic	6.0	155.16
isoleucine	ile, I	$-CH(CH_3)CH_2CH_3$	hydrophobic		131.17
leucine	leu, L	$-CH_2CH(CH_3)_2$	hydrophobic		131.17
lysine	lys, K	$-CH_2CH_2CH_2CH_2NH_2$	basic	10.53	146.19
methionine	met, M	$-CH_2CH_2-S-CH_3$	hydrophobic		149.21
phenylalanine	phe, F	$-CH_2-\Phi*$	aromatic		165.19
proline	pro, P		imino acid (hydrophobic)		115.13
serine	ser, S	$-CH_2OH$	hydrophilic		105.09

CONTINUED ON NEXT PAGE

*Φ is the equivalent of an aromatic ring.

Table 2.1 Continued from previous page.

Amino Acid	Abbrev	Side Chain	Type	pKa	Mol Wt
threonine	thr, T	$-CHOH-CH_3$	hydrophilic		119.12
tryptophan	trp, W		aromatic		204.22
tyrosine	tyr, Y	$-CH_2-\Phi*-OH$	aromatic	10.07	181.19
valine	val, V	$-CH(CH_3)_2$	hydrophobic		117.15

*Φ is the equivalent of an aromatic ring.

α-amino acids. Simple proteins can be further classified based on their amino acid content and water solubility, which will be discussed later.

Conjugated proteins contain, in addition to amino acids, one or more nonprotein groups. Some examples are glycoproteins which contain carbohydrates, lipoproteins which contain lipids, and metalloproteins which contain metal ions.

Proteins are also classified with regard to the three dimensional conformation assumed by the polypeptide chain (i.e., the backbone of the protein molecule). The primary structure of a protein is simply the sequence of the amino acid residues that comprise the particular protein and, as such, is technically not a conformation in space.

The secondary structure of a protein refers to the formation of alpha helix and beta sheet structures. Alpha helices and beta sheet structures are conformations which form spontaneously when it is energetically favorable. These secondary structures arise by hydrogen bonding within the polypeptide chain (Figure 2.3). Random coils formed by a polypeptide are sometimes considered secondary structures although they do not form a predictable structure as do alpha helices and beta sheets.

Two types of proteins that contain secondary structure are referred to as fibrous and globular. Fibrous proteins are extended polypeptide chains that contain either all beta sheet structures or alpha helices. Though there are exceptions, fibrous proteins may contain two or more polypeptide chains, are generally water insoluble, and perform a structural function

A   [1]gly-ile-val-glu-gln-cys-cys-thr-ser-ile-cys-ser-leu-tyr-gln-leu-glu-asn-tyr-cys[20]

B

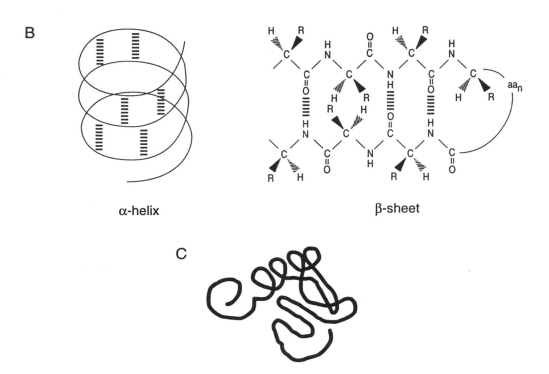

α-helix   β-sheet

C

Figure 2.3   Primary, secondary, and tertiary protein conformations: (A) primary structure—partial sequence of human insulin; (B) α-helix and β-sheet; (C) tertiary structure.

within the organism. Globular proteins, on the other hand, are approximately spherical due to folding and bending of the polypeptide chain. The folding of the protein chain is referred to as tertiary structure, which is defined as the spontaneous folding of the extended polypeptide chain upon itself into a sphere-like, or globular shape (Figure 2.3). Fibrous proteins typically do not contain tertiary structure.

Globular proteins also contain various proportions of beta sheet and helix structures in addition to tertiary structure. The folding of the chains are due to bonding interactions between the side chains of the amino acids within the polypeptide (Figure 2.4).

Tertiary structure can also be stabilized by covalent forces from disulfide bonds, which occur by the oxidation of the thiol side chains present in cysteine residues within the polypeptide chain (Figure 2.5). Disulfide bonds can also be formed between two or more polypeptide chains, for example, between the heavy and light chains in antibodies.

A unique feature of globular proteins is that, due to the folding of the polypeptide, the hydrophilic amino acid residues, i.e., amino acids with a polar (water soluble) side chain, are on the surface of the protein, and the hydrophobic amino acid residues, i.e., amino acids with a nonpolar (water insoluble) side chain, are buried in the interior. The folding of the polypeptide chain in this fashion results in proteins being soluble in water and/or aqueous salt solutions.

Globular proteins can also have quaternary structure. This is when two or more globular proteins, referred to as subunits, associate via noncovalent forces to form larger protein complexes. Proteins that have extensive quaternary structure generally play a multifunctional role in the biochemistry of the cell, e.g., ATPase.

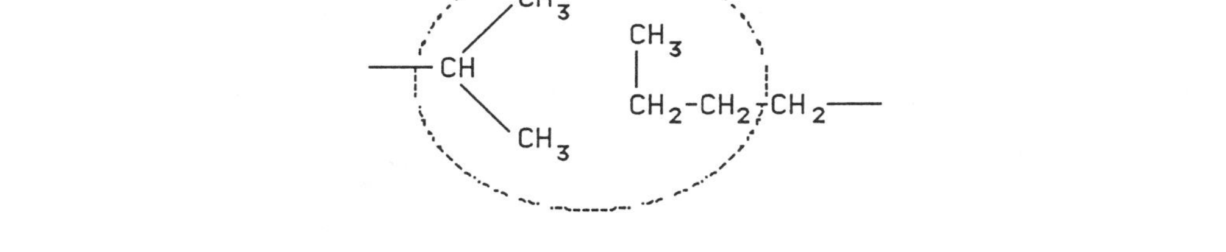

Figure 2.4   Types of bonding responsible for tertiary structure: (A) hydrogen bond; (B) electrostatic interaction (saltbridge); (C) hydrophobic interaction.

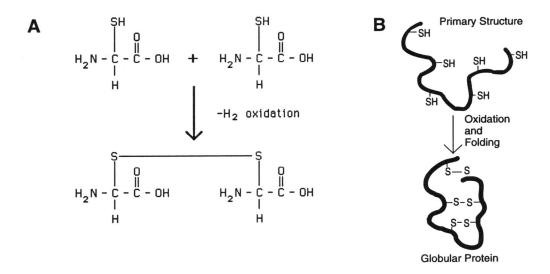

Figure 2.5   Disulfide bond formation: (A) linkage of two cysteines; (B) the effect of disulfide bonds on protein conformation.

Protein Function

Proteins can also be classified by their biological function. Representative examples of the major classes of proteins, including their biological function are shown in Table 2.2.

Characterization of Proteins

Proteins are typically characterized by their molecular weight, amino acid composition and sequence, isoelectric point (pI), hydrophobicity, sedimentation rate, electrophoretic mobility, and biological activity/affinity. The molecular weight of a protein is the mass of one mole of protein, usually measured in units called daltons. One dalton is the atomic mass of one proton or neutron. The molecular weight can be estimated by a number of different methods, including gel filtration, sedimentation rate, electrophoresis, and more recently by mass spectrometry. The amino acid composition is the ratio of the constituent amino acids of the protein while the sequence is the order in which the amino acids are arranged. The isoelectric point (pI) refers to that value of pH where the protein carries a net charge of zero. As you will see, the pI value for a protein is an extremely important parameter to consider when designing a protein purification scheme. Hydrophobicity is a measure of a protein's water solubility and, like pI, is a very useful parameter to consider when planning a purification. (Protein isoelectric point and hydrophobicity determinations will be discussed in more detail later.) Sedimentation rate refers to the speed with which a protein will travel through a viscous medium under the influence of gravity, i.e. centrifugation. Electrophoretic mobility is similar to sedimentation except the protein moves through the viscous medium (i.e., the gel) under the influence of an applied electric field. (Electrophoresis will be discussed in more detail in Chapter 6.) The purification and analysis of proteins generally exploits one or more of the above mentioned properties.

Table 2.2   Classification of Proteins by Function

Function	Description	Example
Enzymes	These proteins catalyze most biological reactions and are the foundation of metabolism.	α-galactosidase
Transport & Storage	Many molecules, such as oxygen, are transported and stored by proteins.	hemoglobin
Coordinated Motion	Cellular and systemic movement are based on protein motion.	actin
Structural	Proteins play an important role in cellular and organismic sructure. Chromosomal scaffolding and hair are two examples.	collagen
Immune Protection	Immunoglobulins are a distinct class of proteins which identify and attack foreign substances.	IgG
Regulatory	Numerous proteins are involved in homeostasis, metabolic regulation, and cellular differentiation.	catabolite activator protein

Measurement of Biological Activity

A more practical parameter used for characterizing proteins is biological activity. A protein activity assay is a measurement of the protein's biological activity as it correlates to protein purity and concentration. Of course, the assumption made here is that the protein has a biological activity that can be measured. The protein activity assay is to be distinguished from another type of analysis referred to as the protein assay or assay for total protein concentration. The assay for total protein concentration simply quantitates the mass of protein present in a sample, based on the chemical or physical properties of the protein.

An example of a protein activity assay (relevant to our purposes in this manual) is the enzyme assay for α-galactosidase. The assay is based on the colorimetric determination of p-nitrophenol, which is released from the synthetic substrate, p-nitrophenyl-α-D-galactoside, by α-galactosidase. An enzyme assay, therefore, is a specific type of protein activity assay.

Biochemical reactions are defined by the rate of the reaction, where reaction rate is defined as the disappearance of substrate, or the appearance of product, over time. The biological function of any enzyme is to accelerate the rate of the associated biochemical reaction. An enzyme assay is a special type of protein activity assay where the increase in reaction rate catalyzed by the enzyme is measured and quantified. This measurement, reflecting the reaction rate increase, correlates with the purity and concentration of the particular enzyme. Simplified, the faster a substrate is consumed or a product accumulates, the greater the concentration and/or purity of the enzyme.

In its basic form, the assay is the quantitative measurement of the rate (product released/time) for the specific biochemical reaction. A useful enzyme assay will be highly sensitive, specific for a given enzymatic activity or substrate with high precision, and reasonably convenient to perform. The assay measures enzyme activity, which is usually expressed in enzyme activity units (U), where one unit is defined as the number of micromoles of product yielded per minute (μmoles/min). The enzyme specific activity, which is a more useful term with regard to monitoring a purification process, is the number of enzyme units per milligram of total protein (U/mg or μmoles/min/mg). The specific activity is used in a purification table and is a reflection of the increase in enzyme activity per mg of total protein (purity) after each step in the purification process. A purification table is a compilation of the purification data that illustrates the effect on enzyme activity after each step in the purification process. Enzyme activity and enzyme specific activity, including the construction of a purification table for α-galactosidase, will be covered in more detail later in this manual.

2.3 EXPERIMENTAL DESIGN AND PROCEDURES

The purification scheme used in this manual is presented as an instructional tool that adequately represents a real world situation within the limited scope of an introductory laboratory manual in biotechnology. The scheme begins with inoculation of yeast extract/peptone based broth (YP) with *Saccharomyces carlsbergensis*. YP broth is a medium rich in nutrients and, with the addition of sugar, provides a standard media for the vigorous growth of yeast. After an incubation period, the turbid culture is analyzed for α-galactosidase activity, and the cellular location of the enzyme must be determined prior to its purification. If the α-galactosidase is found associated with the cells, then the yeast must be lysed so to release the enzyme. If α-galactosidase activity is found in the culture supernatant, then no cell lysis is necessary, and separating the cells from the broth by centrifugation represents the first step in the purification. In either case, the cell lysate or supernatant is referred to as the crude extract.

The α-galactosidase is captured (i.e. isolated) from the crude extract by batch chromatography. This capture step isolates the α-galactosidase from the majority of contaminating proteins and also from any other contaminating biomolecules (DNA, carbohydrates, etc.). Column chromatography is subsequently used to further purify the captured α-galactosidase. This purification step is similar to the batch method, but is more efficient due to the use of the column and salt gradient, the goal being to separate the α-galactosidase from any contaminating proteins that were carried over from the batch purification.

The final step in our scheme uses the technique of electrophoresis for both analysis and purification. Electrophoresis is used to evaluate the previous steps in the purification (batch, column, etc.) by determining the level of purity of the α-galactosidase. The identity of the α-galactosidase is also confirmed by using several analytical detection techniques.

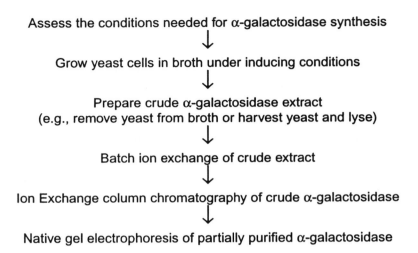

Assess the conditions needed for α-galactosidase synthesis
↓
Grow yeast cells in broth under inducing conditions
↓
Prepare crude α-galactosidase extract
(e.g., remove yeast from broth or harvest yeast and lyse)
↓
Batch ion exchange of crude extract
↓
Ion Exchange column chromatography of crude α-galactosidase
↓
Native gel electrophoresis of partially purified α-galactosidase

Figure 2.6 Purification scheme for α-galactosidase.

This purification scheme represents some of the steps routinely used in the laboratory for protein purification, both in research and industry. In summary, the purification steps are: (1) the production of the protein; (2) release of the protein into a crude extract; (3) a rapid batch process to capture the protein; (4) an efficient method(s) to purify the protein to apparent homogeneity; and finally, (5) analysis of protein purity (Figure 2.6).

The scheme depicted in Figure 2.6 will be performed in the experimental sections of Chapters 2 through 6 of this manual.

Previously, two types of growth media were prepared and inoculated with yeast in order to assess the production of α-galactosidase. These broths will now be analyzed for both protein content and α-galactosidase activity. The activity of the yeast broths are measured to determine which growth medium will yield adequate amounts of crude enzyme for purification. The results of these experiments should show: (1) which medium best induces the synthesis of α-galactosidase; and (2) a comparison of α-galactosidase activity between general and defined media (optional).

Time Course Study of α-Galactosidase Assay

Throughout the purification process of α-galactosidase, enzyme assays will be performed to assess the purification. An important criterion for this assessment is that the enzyme assay is linear, i.e., the substrate is not limiting, and product is produced evenly during the assay incubation period. A time course study is performed to insure that the progress of the reaction is linear over the time scale used in the enzyme assay. The study determines a practical time duration for the assay procedure. The assay will be carried out over a time frame of 30 minutes at constant enzyme concentration during which aliquots of the assay at specified times are quenched, and the activity of the aliquots measured. The activity, or more specifically the absorbance, is measured, and a graph of absorbance versus time is plotted. In this way, you will determine that the enzyme reaction is linear over the time frame used in the assay procedure. If the substrate in this reaction were limiting, i.e., the enzyme consumed the p-nitrophenyl-α-D-galactoside in less than 30 min, the graph would increase and then plateau.

Typically, the enzyme assay is also studied at various enzyme concentrations to insure that the reaction is also linear over the enzyme concentrations used in the assay. The data is evaluated by plotting the absorbance of the assay (a constant incubation time) versus the increasing enzyme concentration. The plot shows over what enzyme concentration the assay is linear, and this information is necessary to determine the limits of enzyme concentration for the assay procedure.

The rationale for the α-galactosidase assay will be described below. However, it is important to verify that an enzyme assay is linear prior to its use. Consequently, this experiment is performed first. Furthermore, due to time constraints, you will not perform the concentration analysis but will assume the reaction is linear over the enzyme concentration range of our assay conditions.

Materials

Yeast cultured on YPG

13 × 100 mm test tubes

0.5 M sodium acetate buffer, pH 4.5

MORE...

0.1 M sodium carbonate solution

0.1 M p-nitrophenyl-α-D-galactoside, in water

Spectrophotometer

Micropipettes

Microfuge tubes

Method

1. Add 1 ml of yeast culture broth to a microfuge tube and pellet the yeast by centrifuging for 1 min at maximum speed. Transfer the supernatant to a separate tube and save.

2. Prepare six test tubes labeled 5, 10, 15, 20, 25, and 30 minutes. To each test tube, add 50 μl of 0.5 M sodium acetate, pH 4.5 buffer, 25 μl of YPG yeast broth supernatent, and 25 μl p-nitrophenyl-α-D-galactoside. The reaction is initiated (time = 0 minutes) when the substrate is added to the reaction mixture. The tubes are sampled at the times designated.

3. The reaction in each tube is quenched at the appropriate time by adding 3 ml of a 0.1 M sodium carbonate solution. If the absorbance of the solution is too high (> 1.0 AU), increase the dilution factor by adding an additional 3 ml of sodium carbonate buffer.

4. The spectrophotometer is zeroed at 410 nm with a blank consisting of 50 μl of acetate buffer, 25 μl of substrate, and 25 μl of water to which 3 ml of the 0.1 M sodium carbonate solution is added. The absorbances of the quenched reactions are then measured.

5. Plot the data with absorbance on the y axis versus time on the x axis. Examine the data to determine over what time frame the enzyme assay is linear with respect to absorbance (i.e., concentration) of the p-nitrophenol assay product.

The standard α-galactosidase assay has an incubation time of five minutes. The activity is proportional to product liberated per minute so that the correlation of absorbance to time should be linear over the 30 min time frame. Does the plotted data confirm this linearity?

Analysis of Culture Broth for α-Galactosidase Activity by the Standard α-Gal Enzyme Assay

The amount of α-galactosidase present in the broth will be determined by an enzyme assay (referred to as the α-gal assay in this manual). This assay is specific for α-galactosidase activity and is insensitive to the presence of contaminating proteins, buffer, salts, etc. This assay works even with crude culture broth containing yeast cells. The α-gal assay relies on the ability of α-galactosidase to hydrolyze the α-glycosidic bond of the synthetic substrate p-nitrophenyl-α-D-galactoside. The substrate is combined with enzyme at pH 4.5, and the hydrolysis of the substrate produces galactose and p-nitrophenol (Figure 2.7).

The assay is quenched (stopped) by addition of a sodium carbonate solution which raises the pH of the solution and stops the reaction. This

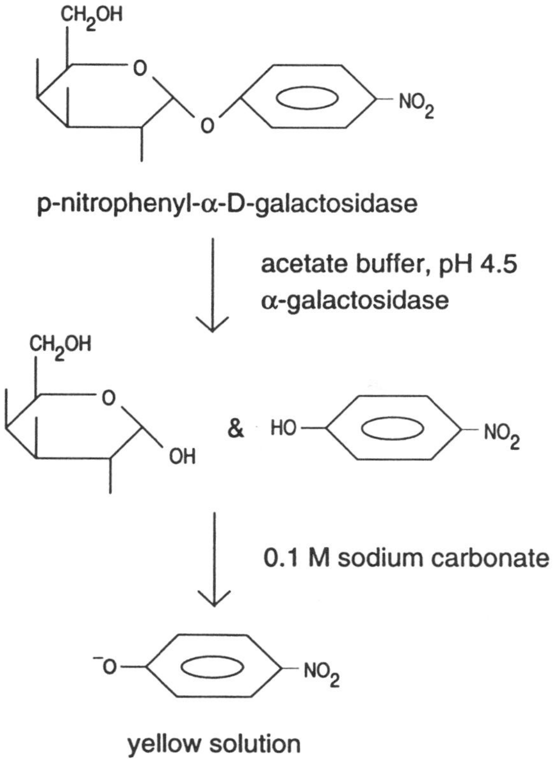

p-nitrophenyl-α-D-galactosidase

acetate buffer, pH 4.5
α-galactosidase

0.1 M sodium carbonate

yellow solution

Figure 2.7 Hydrolysis of p-nitrophenyl-α-D-galactoside by α-galactosidase.

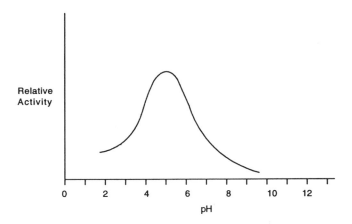

Relative Activity

pH

Figure 2.8 Activity versus pH curve for α-galactosidase.

is because the enzyme activity for α-galactosidase is maximum at pH 4.5 and essentially stops at pH > 8.0 (Figure 2.8). The p-nitrophenol produced by the hydrolysis reaction has an intense yellow color above pH 8.0 and is quantitated by measuring the absorbance of the solution at 410 nanometers with a spectrophotometer.

In the typical α-gal assay, the reaction is incubated at 30°C (or room temperature) for five minutes and then terminated by quenching with sodium carbonate. The activity of the enzyme preparation is then reported in activity units (U) where one activity unit equals 1 μmole of product released per minute.

For this assay, the concentration of p-nitrophenol can be determined spectrophotometrically and by applying Beer's Law. Beer's Law states that $A = \varepsilon bc$, where "**A**" is absorbance (i.e., the OD), "**ε**" is the molar extinction coefficient ($M^{-1}cm^{-1}$), "**b**" is the length of the light path through the sample or cuvette width (cm), and "**c**" is the concentration of the absorbing solute (M). The molar extinction coefficient (ε) is a value specific for each molecule at a specific wavelength and it is used to equalize the two sides of the equation. For p-nitrophenol, ε is 18,300 $M^{-1}cm^{-1}$ at 410 nm. Therefore, activity per ml is calculated using Equation 2.1.

Equation 2.1   Calculation of α-galactosidase activity per ml

Objective: Determine the number of units of α-galactosidase.

Activity (U) = μmoles p-nitrophenol liberated/min.

First, calculate concentration of p-nitrophenol (pNP) by rearranging $A = \varepsilon bc$,

$$c = A/\varepsilon b$$

Second, to calculate the number of μmoles pNP, convert from Molar to μmoles and adjust for the 3 ml sample volume,

μmoles p-nitrophenol = $(A/b\varepsilon) \times 0.003 \, l \times (1 \times 10^6 \, \text{μmoles/mole})$.

Simplified,

μmoles p-nitrophenol = $0.164 \, \text{μmoles} \times A$.

Third,

Activity (U) = $(0.164 \, \text{μmoles} \times A)/5 \, \text{min}$.

Finally, the number of units per ml of enzyme sample is calculated by dividing U by the original assay sample volume (25 μl or 0.025 ml):

U/ml = U/0.025 ml = $[(0.164 \, \text{μmoles} \times A)/5 \, \text{min}]/0.025 \, \text{ml}$.

Materials

13 × 100 mm test tubes
0.5 M sodium acetate buffer, pH 4.5
0.1 M p-nitrophenyl-α-D-galactoside, in water (aliquot and freeze)
Constant temperature block or water bath
0.1 M sodium carbonate (Na_2CO_3)
Micropipettes and tips
Microfuge tubes
YPD and YPG culture broths
Marking pen
Spectrophotometer—one which accepts 13 × 100 mm tubes

Method

1. The following steps define the α-galactosidase assay that will be used repeatedly throughout this manual. The steps are essentially the same whether the sample being analyzed is culture broth, supernatant, or purified enzyme.

2. For each culture broth, add the following to a 13×100 mm test tube:

   50 μl of 0.50 M sodium acetate buffer, pH 4.5

   25 μl of yeast culture broth

   25 μl of p-nitrophenyl-α-D-galactoside substrate

3. The reactions are incubated at 30°C in a dry block or water bath (or at room temperature if neither is available) for 5 min.

4. Enzyme activity is quenched by adding 3 ml of a 0.1 M sodium carbonate solution to the assay solutions. The appearance of a yellow color due to the production of p-nitrophenol indicates the presence of enzyme activity. Optimum enzyme activity is at pH 4.5 with activity essentially zero above pH 8.0. Prior to analysis of the assay solution, the spectrophotometer is zeroed at 410 nm with a blank consisting of 50 μl of acetate buffer, 25 μl of substrate, 25 μl of water (or buffer), and then 3 ml of carbonate buffer.

5. The absorbance of the quenched assay solutions are then measured at 410 nm and the activity per ml is calculated by using Equation 2.1.

6. Based on the observed activity of the yeast culture broth, choose which medium is best for the large scale production of α-galactosidase.

Colorimetric Protein Assays

The total protein concentration, which includes the enzyme of interest along with all other contaminating proteins, must be known in order to determine specific activity. Total protein concentration is usually determined by absorbance of the protein at 280 nm or by one of several colorimetric protein assays.

Ultraviolet absorption at 280 nm is the simplest method for determination of protein concentration and has several advantages over the colorimetric methods, the most important being that the method is not destructive to the protein sample. The sample is simply added to a quartz cuvette and the absorption at 280 nm is measured.

Like p-nitrophenol, each species of protein has a unique extinction coefficient (ε). If ε is known and the protein is in a pure solution, then calculating the concentration simply involves applying Beer's Law. However, it is difficult to determine ε for an unknown protein. A variation of Beer's Law is used to determine protein concentration in mg/ml. By substituting "absorption constant" for "extinction coefficient," a similar relationship between protein concentration and absorbance is defined by Equation 2.2.

Equation 2.2

$$\text{Absorbance} = (\text{absorption constant})\ \text{ml mg}^{-1}\text{cm}^{-1} \times \text{Concentration} \times 1\ \text{cm}$$

therefore:

$$\text{Absorbance}/\text{absorption constant} = \text{concentration (mg/ml)}$$

The absorption constant is specific for the particular protein being measured and is determined using a cuvette with a 1 cm path length containing a protein solution of known concentration (in mg/ml). When measuring a solution of protein whose absorption constant is unknown, or when measuring a protein solution consisting of a complex mixture of proteins, Equation 2.3 is used.

Equation 2.3

$$\text{Concentration (mg/ml)} = \text{absorbance of protein at 280 nm}$$

The above equation assumes a path length of 1 cm and an absorption constant of unity (i.e., 1 OD = 1 mg/ml protein). This measurement is considered qualitative since a true molar extinction coefficient or absorption constant is not used. However the method is useful to approximate

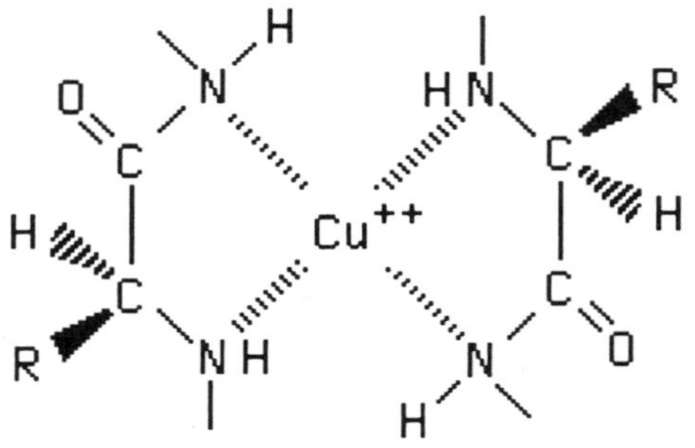

Figure 2.9 Structure of Biuret protein complex.

protein concentrations because it is simple and nondestructive. A further refinement of this method by Warburg and Christian (1941) corrects any errors in the absorbance value due to the presence of nucleic acids in the sample.

An alternative approach to protein concentration measurement is by using a colorimetric assay. There are essentially three popular colorimetric methods for protein determination: (1) Biuret (Gornall et al, 1949); (2) Lowry (Lowry et al, 1951); and (3) Bradford (1976).

The Biuret and Lowry methods are similar in that the reaction used to generate color is specific for the presence of peptide bonds. The Biuret reagent consists of copper sulfate and sodium potassium tartrate mixed together in a solution of sodium hydroxide. Potassium iodide is added for stability. The Cu ion forms a square planar complex between the four amide nitrogen atoms of two parallel polypeptide chains in the protein molecule, the result being a blue protein solution (Figure 2.9). The absorbance measured at 545 nm is proportional to protein concentration when compared to a standard curve generated using proteins of known concentrations.

The Lowry is similar to the Biuret method in that copper ions which interact with the peptide bond are detected. However, color development is dependent upon the reduction of a phosphomolybdic–phosphotungstic reagent (i.e., Folin & Ciocalteu's Phenol Reagent) by the copper–protein complex. Protein causes the Lowry reaction to turn from yellow to dark blue/purple.

The Bradford method of protein determination is based on the binding of a dye, Coomassie Blue G, to the protein. This binding shifts the absorption maximum of the dye from red to blue. A solution of Coomassie Blue

G dye is mixed with phosphoric acid and ethanol. The dye reagent is mixed with the protein sample, and the presence of protein is indicated by the formation of a deep blue color. The absorbance of the solution is measured at 595 nm and is proportional to protein concentration when compared to a standard curve. The Bradford method does not measure the presence of peptide bonds but detects specific amino acids, such as arginine, which is believed to be responsible for the binding of the dye to the protein. A standard curve will be generated for the Bradford reaction in this chapter.

All three methods require the construction of a standard curve using known concentrations of a protein such as BSA or gamma globulin. The Bradford and Lowry methods are the most sensitive with respect to protein mass and can detect proteins on a µg/ml scale. The Biuret method is the simplest to perform and is the least sensitive to errors from experimental artifacts (i.e., substances such as buffer salts that can interfere with the color generating process).

Biuret Protein Assay

Materials

> Copper sulfate, pentahydrate
> Sodium potassium tartrate
> Sodium hydroxide solution, 2.5 M
> Potassium iodide
> Bovine Serum Albumin (BSA)
> 13 × 100 mm test tubes
> Spectrophotometer
> Bovine Serum Albumin sample, unknown concentration
> Micropipettes and tips

Method—Preparation of the Biuret Reagent

1. Add 1.5 grams of copper sulfate, pentahydrate, and 6 grams of sodium potassium tartrate to a 1 liter beaker. Add 500 ml of distilled

water and 300 ml of 2.5 M sodium hydroxide (carbonate free) to the beaker and stir the solution until dissolved.

2. Add 1 gm of potassium iodide to the solution and dilute to 1 liter with distilled water. Stir until all solids are dissolved.

3. The reagent is placed in a plastic bottle and is stable indefinitely when stored at 4°C. The freshly prepared Biuret reagent will be used to generate a standard curve that will then be used for protein assays in later experiments.

Method—Protein Determination by the Biuret Method

1. Prepare a 10 mg/ml solution of Bovine Serum Albumin (BSA) by dissolving 200 mg of BSA in 20 ml of water. This solution can be stored at −20°C.

2. Label ten 13 × 100 mm tubes (1–10) and add 1.0, 0.8, 0.6, 0.4, 0.2, and 0.1 ml of the 10 mg/ml BSA solution into tubes 1–6, respectively. No protein is added to the seventh tube which will serve as the blank. Tubes 8–10 are reserved for an unknown protein solution which will be analyzed (in triplicate) to demonstrate the method.

3. Bring the volume of the solutions in tubes 1–7 to 1 ml with distilled water.

4. Add 4 ml of Biuret reagent to each tube, mix, and allow the solution to incubate for 20 minutes.

5. Zero the spectrophotometer using the blank (0 mg/ml) at 545 nm and then measure the absorbance of each solution.

6. Plot the data as absorbance on the y axis against mg protein on the x axis. For example, tube #2 contains 0.8 ml of a 10 mg/ml solution which equals 8 mg BSA.

7. Add 1 ml of the unknown protein solution to tubes 8–10 and perform the Biuret assay as described above. Measure the absorbance of the unknown solution. Determine where the absorbance falls on the standard curve and estimate the mg protein content of the un-

known. The concentration of the original protein solution is given by the following equation:

$$\text{concentration of unknown} =$$
$$\text{mg of protein determined by assay/ml unknown sample}$$

Lowry Protein Assay

Materials

Reagent A—2 % sodium carbonate in 0.1 M sodium hydroxide solution

Reagent B1—0.5% copper sulfate pentahydrate solution

Reagent B2—1% sodium tartrate solution

Reagent C—combine solutions A, B1 and B2 in a ratio of 100:1:1

1 N Folin & Ciocalteu's Phenol Reagent—diluted with water from 2 N stock solution (Sigma F 9252)

13 × 100 mm test tubes

Spectrophotometer

Method

1. Prepare a 0.5 mg/ml working solution of BSA by diluting the 10 mg/ml BSA stock 1:20 with deionized water. Excess diluted BSA may be stored at –20°C. To prepare a standard curve, add 400 µl, 300 µl, 200 µl, 100 µl, 50 µl, and 25 µl of the diluted BSA solution to tubes 1 through 6. Respectively, add 0 µl, 100 µl, 200 µl, 300 µl, 350 µl, and 375 µl of deionized water to each tube to bring the total volume up to 400 µl.

2. In triplicate, add 400 µl of unknown protein samples to 13 × 100 mm test tubes (tubes 7–9). Add 400 µl water to tube 10.

3. Add 2 ml of Reagent C to each tube and incubate for 10 minutes.

4. Add 200 µl of 1 N Folin & Ciocalteu's Phenol Reagent to each tube. The Folin & Ciocalteu's Phenol Reagent reagent is unstable, and the mixture must be vigorously mixed as it is being added. Let the tubes incubate for exactly 30 minutes.

5. After 30 minutes, zero the spectrophotometer using tube 10 at 550 nm and then measure the absorbance of each tube.

6. For the protein standards (tubes 1–6), plot the absorbance on the y axis against mg protein on the x axis. Determine where the absorbance falls on the standard curve and the corresponding protein mass of the unknown samples.

Note: When assaying solutions containing Tris buffer, as a control, analyze the Tris buffer by the Lowry assay and subtract the value from the protein sample.

Bradford Protein Assay

Materials

Coomassie Blue G

Ethanol

Phosphoric acid, 85%—Caution! Phosphoric acid can cause burns!

13 × 100 mm test tubes

Bovine Serum Albumin

Spectrophotometer

Micropipettes and tips

Method—Preparation of the 5X Bradford Reagent

1. Dissolve 100 mg of Coomassie Blue G in 50 ml of ethanol and 100 ml of phosphoric acid (85%). The mixture is stirred for approximately 10 minutes. The solution is diluted to 200 ml with distilled water and filtered.

Method—Protein Determination by the Bradford Method

1. A 10 ml aliquot of the Bradford concentrate solution is diluted 1:5 (1×) with water to prepare the working solution.

2. In separate tubes, prepare Bovine Serum Albumin dilutions of 1.2, 1.0, 0.8, 0.6, 0.3 and 0.1 mg/ml from the 10 mg/ml stock.

3. Label six 13 × 100 mm test tubes 1–6, a seventh tube as "blank" and tubes 8–10 as "unknown."

4. To tubes 1–6, add 100 µl of the 1.2, 1.0, 0.8, 0.6, 0.3 and 0.1 mg/ml BSA solutions, respectively, and 100 µl of distilled water is added to the blank tube. Analyze in triplicate 100 µl of the unknown sample of BSA in tubes 8–10.

5. Add five (5) ml of diluted (1×) Bradford reagent to each of the 10 tubes. Stir and let incubate for 20 minutes.

6. The spectrophotometer is set at 595 nm, and the instrument is zeroed against the water blank. The absorbance of the other tubes is then measured at 595 nm. The absorbance of each tube (y-axis) is plotted against the protein mass in µg (x-axis) and a straight line is drawn through the points.

7. The amount of protein in the unknown tubes is determined by its absorbance and comparison to the standard protein curve. The concentration of the solution is calculated by the following equation:

$$\text{Concentration of unknown} = \text{mg of Protein determined by assay}/0.1 \text{ ml}$$

Preparation and Inoculation of YP Broth for α-Galactosidase Production

The analysis of the YPD and YPG culture broths should provide clear data as to which medium is preferred for the production of α-galactosidase from *S. carlsbergensis*. Using that data and the experiences of preparing and inoculating media, select and prepare a broth, and then culture *S. carlsbergensis* for the production of α-galactosidase. The purification protocols in the following chapters require that you start with a least 50 ml of culture broth. Thus, you need to prepare an appropriately sized culture, using the preferred medium. It is suggested that the culture be incubated for 48 hours so that adequate amounts of enzyme are produced. If necessary, the culture broth can be refrigerated until needed.

STUDY QUESTIONS

1. In the case of a low absorbance reading for the α-galactosidase assay due to low enzyme concentration, what experimental conditions of the assay can be changed to improve the absorbance reading?

2. What is the primary advantage of using synthetic culture media over the yeast extract, peptone media when isolating proteins?

3. What buffer contaminants interfere with the Bradford reagent? (Hint: Read the corresponding references at the end of the chapter.)

4. How could the absorption constant for an unknown protein be determined experimentally?

FURTHER READINGS

Stryer L (1988): *Biochemistry*, 3rd Ed. New York: WH Freeman

REFERENCES

Bradford M (1976): A rapid and sensitive method for the quantitation of microgram quantities of protein utilizing the principle of protein–dye binding. *Anal Biochem* 72:248–254

Gornall A, Bardawill C, Maxima D (1949): Determination of serum proteins by means of the Biuret reaction. *J Biol Chem* 177:751–766

Lazo PS, Ochoa AG, Jascón S (1977): α-Galactosidase from *Saccharomyces carlsbergensis*: Cellular localization, and purification of the external enzyme. *Eur J Biochem* 77:375–382

Lazo, PS, Ochoa AG, Jascón S (1978): α-Galactosidase (melibiase) from *Saccharomyces carlsbergensis*: Structural and kinetic properties. *Arch Biochem Biophys* 191:316–324

Lowry H, Rosebrough N, Farr A, Randall R (1951): Protein measurement with the Folin Phenol Reagent. *J Biol Chem* 193:265–275

Warburg O, Christian W (1941): Isolierung und kristallisation des gärungsferments enolase. *Biochem Z* 310:384–421

3

Protein Isolation and Preparation of Crude Extract

3.1 OVERVIEW

The initial step in protein isolation from its source is to physically or chemically disrupt the biological tissue or organism in order to release the protein into the extract. In some cases, the organism secretes the protein, and cell disruption is unnecessary. The protein is usually produced in dilute concentration so it is advantageous if the initial isolation also results in the concentration of the protein. An important feature in any isolation procedure is product yield with the goal being to extract as much of the desired protein as possible in a minimum amount of time.

Depending on the source of the protein, the means by which a crude extract is produced will differ. The background section of this chapter will cover the salient features of cell disruption and crude extract preparation, including maintenance of protein stability.

In Chapter 2, glucose and galactose were assessed for their ability to induce the synthesis of α-galactosidase. Galactose was clearly shown to induce enzyme synthesis to a greater extent than glucose. The batch cultivation of *S. carlsbergensis* on galactose based YPG will yield sufficient enzyme for its subsequent purification. The next step in the purification is to isolate the enzyme from its biological source and to stabilize the resulting crude extract.

The experiments for this chapter will be to determine the distribution of the α-galactosidase activity between the yeast culture supernatant, periplasm, and cytoplasm. The specific activity of the fraction with the highest enzyme activity will be determined and used as a measurement of enzyme purity.

3.2   BACKGROUND

The isolation of proteins from natural sources is a complicated and delicate task that cannot be distilled down to a set of simple instructions. The differences between protein species in their amino acid composition and sequence creates variation in their physical and chemical characteristics which are exploited for purification. The enormous variation in these properties precludes the use of standardized purification protocols. The following list, however, delineates crucial factors that must be addressed in any protein isolation and purification. These factors are:

(1) development of an assay for protein activity;

(2) choice of a suitable protein source (including availability);

(3) isolate the protein of interest from either extracellular sources (e.g., supernatant) or cell mass (e.g., cell lysate);

(4) develop conditions to maintain protein stability in the crude extract;

(5) subsequent purification and stabilization (multiple steps) of the desired protein; and

(6) analysis for structure determination and/or purity.

The first factor was previously covered in Chapter 2. In this chapter we will focus on (2) and (3) with the remaining factors covered later.

The purity of a given protein is defined by its chemical purity and its biological activity (i.e., biological purity). The chemical purity of a protein is determined by such techniques as amino acid analysis, sequence analysis, etc. The biological purity of a protein is generally measured by its specific activity which is the biological activity (as defined in Chapter 2) divided by the total mass of protein in milligrams. These measurements in purity are used to assess the efficiency of the purification process.

In the purification process, the protein is effectively concentrated through a series of steps, concurrent with the removal of contaminating biomolecules. The result of this purification is increased biological activity and purity within a given volume, while total protein concentration decreases. The assay for protein activity is used to determine the efficiency of each step in the overall purification of the protein. Therefore, it is important to assay the crude cell or tissue extract at the beginning of the purification procedure in order to get a baseline value of protein activity.

Selection of Protein Source

Selection of an appropriate protein source is a fundamental aspect of purification. In many cases, the source of the protein or enzyme is either limited or predetermined and the researcher must make do with what is available. When the freedom of choice presents itself, it is best to choose a source where the desired protein represents a large percentage of the total protein present. Typical sources used for protein production are bacteria, yeast, and cell culture, which can be genetically and/or physiologically altered to increase the level of the desired protein. Tissues, organs, and at times whole organisms can also be used as a source.

Cell Disruption Techniques

Most proteins are cell associated (i.e., not extracellular) and must be released from the biological source through some type of disruption followed by solubilization into an appropriate buffer. The disruption process will vary from gentle (e.g., cell lysis) to severe (e.g., ultrasonication) depending on the type of cell or tissue that is the source of the protein. A detailed discussion of the many techniques of cellular disruption is beyond the scope of this book, however, a summary of the more common methods is listed in Table 3.1.

Table 3.1   Common Methods of Cell Disruption

Technique	Target for Disruption
Osmotic cell lysis	cellular membranes
Enzymatic cell lysis	prokaryotes
Autolysis/chemical disruption	yeast
Homogenization and grinding	animal and plant cells/tissues
Agitation with abrasives (bead mill)	prokaryotes, yeast, molds
Shear technique (French press or sonication)	prokaryotes, yeast, molds

Prokaryotes (i.e., bacteria) can be lysed either by mechanical means or by treatment with lytic enzymes. Enzymatic cell lysis is a gentle and selective method that minimizes protein denaturation but a major disadvantage of this technique is that the introduction of lytic enzymes may interfere with subsequent protein purification. In a typical procedure, the bacterial cell wall is degraded with enzymes, and the cell membrane is lysed by detergent or osmotic pressure.

Mechanical methods of prokaryotic cell lysis include both liquid shear and agitation with abrasives. The mechanical methods for cell lysis are easily scaled up and do not require the addition of any potentially contaminating substances.

Agitation with abrasives typically uses a bead mill where rotating glass beads disrupt the cell through high shear and cellular impact. The French press is a typical example of lysis by liquid shear where a suspension of cells are compressed and then rapidly depressurized, resulting in cell lysis. Sonication is another popular liquid shear technique for cell lysis but is limited to small scale operation.

As with prokaryotes, yeast cell lysis employs both enzymatic and mechanical methods. The yeast cell wall, however, is stronger than the prokaryotic cell wall and is not disrupted by sonication. Autolysis is a simple method of yeast cell disruption. The yeast suspension is treated with toluene and incubated for 24 to 48 hours during which time the toluene disrupts the yeast cell membrane and releases hydrolytic enzymes that eventually degrade the yeast cell wall. The principle drawback here is the potential loss of activity of the desired protein due to the indiscriminate release of lytic enzymes by cell membrane destruction.

The French press can be used to disrupt yeast cells but is generally limited to small volumes of cell suspension and require high pressures. The most prevalent mechanical method of yeast cell disruption is by agitation with glass beads. The yield is high, and the process can be easily scaled up to handle liters of cell suspension.

The use of lytic enzymes to disrupt the yeast cell wall is a popular method, but, as with prokaryotes, the technique has problems with extract contamination. Enzymatic digestion, however, is very effective when dealing with the isolation of large enzyme complexes where shear forces may result in denaturation and loss of activity.

The type of cell lysis method used for higher eukaryotes generally depends on the source of the cellular material. Tissue culture cells and animal cellular suspensions are typically swollen in a hypotonic buffer and then lysed using a hand-held homoginizer. A blender is used when dealing with solid tissue, such as large animal organs, due to the possible presence of blood vessels and strong connective tissue. Passing the tissue through a meat grinder prior to processing in the blender facilitates the homogenization process when using tissue with large quantities of skeletal muscle.

A more gentle method of cell disruption utilizes incubation of cells in a glycerol buffer followed by treatment with a hypotonic buffer. The glycerol method is gentle and helps to preserve the activity of large protein complexes. However, the drawback to this method is that it is appropriate only for cell suspensions.

Factors Affecting Activity in the Crude Extract

Proteins are susceptible to denaturation and possible loss of activity after being liberated from their natural environment. Once the protein source is disrupted, the proteins and other biomolecules and cellular debris (assuming cellular disruption) are solubilized in an appropriate buffer. The buffer must provide a stable environment that maintains both the biological and chemical integrity of the protein. Any insoluble material at this point can be removed by centrifugation. Additionally, the buffer composition should be evaluated at every stage in the purification process because, as the protein activity is increased by repeated purification, the requirement for stabilization of the protein by the buffer usually takes on even greater importance.

Several factors have to be considered in the choice of a buffer to insure complete solubilization and stabilization of the protein. The pH of the buffer is extremely important in providing a stable medium for the protein. The optimum pH for protein stability must be independently determined since it may not necessarily coincide with the optimum pH for protein biological activity. An example of this is α-galactosidase which is most stable at pH 7.5 but exhibits optimum enzyme activity at pH 4.5. The buffer itself must be inert so it does not react in any adverse way with the protein being purified.

Proteins containing cysteine may possess free thiol groups. In the presence of another cysteine residue (or some other thiol containing molecule), the free thiols can oxidize and form a disulfide bond. This reaction can lead to loss of activity, especially when the thiol group is part of an enzyme's active site. The reducing environment within the cell may be important for the viability of a protein. When liberated from the cell into a crude extract, however, a protein may oxidize and lose activity. Sulfhydryl reagents, such as 2-mercaptoethanol, dithiothreitol or reduced glutathione, when added to the buffer, simulate the reduced intracellular environment and stabilize the protein preparation.

Heavy metal ion contamination of the buffer system may have deleterious effects on protein activity, especially with proteins containing free thiol groups. The sources of metal ion contamination most commonly originate in the commercial buffer and salt reagents or in the water used to prepare the buffers. The sulfhydryl reagents mentioned above will help to neutralize the effects of metal ions, but a more common and useful reagent to add to the buffer system is ethylenediaminetetraacetic acid (EDTA). The EDTA chelates (binds) divalent metal ions that may contaminate the buffer and renders the ions harmless.

Temperature is an important factor in protein stability because, as with pH, a protein usually has separate optimum temperatures for activity and stability. Generally, the colder the environment the more stable the protein (especially if proteases are present), but there are exceptions to this general rule, and the optimum temperature for stability and activity should be independently determined. Though low temperatures generally enhance protein stability, the cold can negatively affect steps in the purification process. Column chromatography is by far the technique of choice for protein purification and is extremely sensitive to temperature; therefore, choosing the optimum temperature to balance protein stability with purification efficiency is critically important.

The polarity, or hydrophobicity of the buffer system must also be evaluated. The hydrophobic environment is especially important for stabilizing membrane proteins, but it is also an important factor in the preparation of most intracellular proteins that contain appreciable numbers of hydrophobic residues on their surface. The hydrophobicity of the buffer can be increased by adding sucrose or glycerol, usually around 10% v/v of the system but not more than 20%. The amount of glycerol or sucrose that is added should be carefully determined because the increase in hydrophobicity afforded by these reagents will cause a corresponding increase in the buffer viscosity. The disadvantage of reagents such as glycerol is that the increased viscosity of the buffer also has a negative effect on the efficiency of the chromatographic techniques that will be used later to purify the protein. In some cases, the addition of cofactors, lipids, or even substrate can help to maintain protein activity.

The initial cell disruption may release proteolytic enzymes, called proteases into the crude extract. A protease is an enzyme that hydrolyzes the peptide bond of the protein chain, thereby degrading the protein. Due to the possible presence of these protease contaminants, a purification scheme should be executed rapidly so that a minimum of activity is lost from the protein extract. Protease inhibitors, such as diisopropylfluorophosphate (DIFP), phenylmethylsulfonyl fluoride (PMSF), leupeptin, or phosphoramidon can be added to the buffer to render the proteases harmless. Care must be exercised in using these inhibitors to make sure that they have no adverse reactivity with the desired protein product. When using protease inhibitors, handle them carefully since many protease inhibitors are exceedingly toxic.

3.3 EXPERIMENTAL DESIGN AND PROCEDURES

The experiments for this chapter will consist of determining the distribution of α-galactosidase activity in the yeast culture. There are several distinct locations where the enzyme may be localized. First, α-galactosidase may be secreted from the cell, in which case activity will be associated with the supernatant. Alternatively, activity may be associated with the yeast cells. Cell associated enzyme activity can further be localized to the cell wall and periplasm, or to the cell membrane and cytoplasm. Accordingly, a strategy will be adopted for analyzing enzyme activity in the supernatant, whole yeast cells, and protoplasts (yeast with the cell wall removed) from a culture of *S. carlsbergensis* (cultured previously). The purpose of this experiment is to determine whether the supernatant can be used as the enzyme source or if the yeast cells must be disrupted to extract the enzyme. A table will be constructed that will show the distribution of α-galactosidase activity present in the yeast culture supernatent, yeast cells and protoplasts.

α-Gal Assay of Yeast Culture Broth

The α-galactosidase activity of the culture broth is determined prior to the separation of the cells from the supernatant. The measurement allows us to determine how much activity is recovered in the subsequent preparation of yeast culture broth supernatant.

Materials

13 × 100 mm test tubes
0.5 M sodium acetate buffer, pH 4.5
0.1 M p-nitrophenyl-α-D-galactoside, in water
Constant temperature block or water bath
0.1 M sodium carbonate solution
Spectrophotometer
Micropipettes and tips
Yeast culture broth

Method

1. Add 50 µl of 0.5 M sodium acetate, pH 4.5 buffer, 25 µl of the culture broth (supernatent and cells) and 25 µl of the p-nitrophenyl-α-D-galactoside to a 13 × 100 mm test tube. Incubate the assay mixture for five minutes at 30°C (or at room temperature if necessary).

2. Quench the assay by adding 3 ml of 0.1 M sodium carbonate solution. Zero the spectrophotometer with a negative control and then measure the absorbance of the solution at 410 nm.

3. Calculate the activity/ml of the culture broth as described in Chapter 2.3 and enter the value in your notebook.

α-Gal Assay of the Yeast Supernatant

This experiment will involve separating supernatant from the cells of the *S. carlsbergensis* culture and measuring the supernatant α-galactosidase activity. The supernatant is prepared by centrifugation of the yeast culture broth followed by decanting the supernatant from the yeast cell pellet.

Materials

13 × 100 mm test tubes
0.5 M sodium acetate buffer, pH 4.5
0.1 M p-nitrophenyl-α-D-galactoside, in water
Constant temperature block or water bath
0.1 M sodium carbonate buffer
Spectrophotometer
Micropipettes and tips
Centrifuge tubes
Centrifuge

Method

1. Centrifuge 5 ml of the yeast culture broth at 5000 rpm for five minutes. Decant and collect the cell supernatant. Record the volume collected and save the cell pellet for the following experiment. Keep the remaining culture broth cold while not in use.

2. Remove 25 μl of the supernatant and analyze by the α-galactosidase assay as described previously.

3. Calculate the activity/ml of the supernatant. The activity/ml is obtained by dividing the number of activity units by 0.025 ml, which is the volume of the sample used in the assay. Record the value of the enzyme units/ml and the total number of units.

4. The recovery of enzyme activity in the supernatant is obtained by dividing the activity of the yeast supernatant by the activity of the yeast culture (same sample volume used in both) and multiply by 100.

Preparation of Yeast Protoplasts

The protoplasts, which are the yeast cells enzymatically stripped of their cell walls, will be produced, and the amount of α-galactosidase activity liberated during protoplast formation will be determined. The α-galactosidase liberated during protoplast formation represents enzyme localized in the periplasm between the cell membrane and cell wall. The amount of α-galactosidase activity associated with the cell wall and protoplast will be determined and compared with the enzyme activity in the supernatant.

Materials

Yeast pellet (prepared earlier)

Microcentrifuge tubes

Test tube rack

Micropipettes and tips

Centrifuge

50 mM Tris-HCl, pH 7.5

50 mM Tris-HCl, pH 9.5, 2% 2-mercaptoethanol

1.2 M sorbitol, 50 mM Tris-HCl, pH 7.5,

Lyticase, 500 U/ml in 50 mM Tris-HCl, pH 7.5 (Sigma L 8012)

Microscope

Sodium dodecylsulfate (SDS), 10% solution

13 × 100 mm test tubes

0.5 M sodium acetate, pH 4.5

0.1 M p-nitrophenyl-α-D-galactoside, in water

Constant temperature heating block or water bath

0.1 M sodium carbonate

Spectrophotometer

Method

1. Suspend the cell pellet (prepared earlier from 5 ml of yeast culture) in 1 ml of 50 mM Tris-HCl, pH 7.5, spin the tube at 5000 × g for 1 minute to pellet the cells, and decant the supernatant. Repeat this step two more times to ensure removal of extracellular α-galactosidase.

2. Suspend the cells in 1 ml of 50 mM Tris, pH 9.5, 2% 2-mercapto-ethanol. Incubate for 10 minutes at room temperature. Centrifuge and decant the supernatant.

3. Resuspend the cells in 800 μl of 1.2 M sorbitol, 50 mM Tris-HCl, pH 7.5.

4. Add 200 μl of Lyticase (500 U/ml in 50 mM Tris-HCl, pH 7.5). Place the cells on a rocker and incubate at 37°C for one hour.

5. The suspension is examined microscopically to ensure protoplast formation. Phase contrast microscopy will provide the best comparison. Protoplasts are not as easy to detect as the so-called ghosts (the cell wall without the protoplast). Under a bright field microscope, ghosts are barely visible and appear as yeasts but are not refractile (bright). Under a phase microscope, the ghosts appear as dark grey or black yeast cells. The protoplasts appear as refractile, perfectly spherical cells. The addition of SDS to an aliquot will demonstrate lysis by the disappearance of the protoplasts. Pellet the protoplasts by centrifugation and collect the supernatant. Save the protoplasts on ice for the following experiment.

6. Analyze the protoplast supernatant with the α-Gal assay and calculate the activity/ml. Multiply the activity/ml by the total volume of protoplast supernatant to determine the total activity units and record both these values in your notebook.

Lysis of the Yeast Protoplast Cells

The protoplasts are lysed, and the lysate is collected and analyzed for α-galactosidase activity.

Materials

100 mM Tris-HCl, pH 7.5, 100 mM EDTA, 150 mM NaCl

100 mM Tris-HCl, pH 7.5, 100 mM EDTA, 150 mM NaCl, 2% SDS

Sodium dodecylsulfate (SDS), 10% solution

Microscope

Bench top centrifuge

Micropipettes and tips

0.5 M sodium acetate, pH 4.5

0.1 M p-nitrophenyl-α-D-galactoside, in water

13 × 100 mm test tubes

Constant temperature heating block or water bath

0.1 M sodium carbonate buffer

Spectrophotometer

Method

1. Suspend the protoplasts (prepared earlier) in 1 ml of 100 mM Tris, pH 7.5, 100 mM EDTA, 150 mM NaCl (lysis buffer).

2. Add 1 ml of lysis buffer with SDS (100 mM Tris-HCl, pH 7.5, 100 mM EDTA, 150 mM NaCl, 2% SDS). Mix and incubate at 30°C for 30 minutes. Check the cells microscopically for lysis. The presence of 2% SDS does not denature α-galactosidase, therefore activity of the enzyme is retained.

3. Centrifuge the suspension and decant the supernatant. The supernatant contains the α-galactosidase activity released from the yeast protoplasts. Assay the supernatant with the α-gal assay and determine the activity/ml of the solution.

4. Prepare a table (as shown on the next page) comparing the amount of enzyme activity as to its location. Multiply the activity/ml by the corresponding total volume to obtain the total number of activity units. These total activities are a measurement of the amount of enzyme in each fraction.

	Activity (U)/ml	Total Activity (U)
Yeast Culture		
Yeast Supernatant		
Periplasm (released)		
Lysed Protoplasts		

Collection of Supernatant and Determination of Total Protein

The determination of the distribution of α-galactosidase (previous experiment) is necessary prior to the isolation of the enzyme. In situations where one fraction has significantly more activity than others, then that fraction is chosen. If the activities of two or more fractions are similar, then the fraction which allows for the most convenient purification is used. Supernatants are generally cleaner than periplasmic fractions, which are cleaner than cellular fractions.

In this experiment, the supernatant will be used as the source of α-galactosidase for the purification. In order to measure the efficiency of the purification, the amount of total protein in the supernatant is determined by the Biuret method. This value will then be used to determine the specific activity of the yeast supernatant. The specific activity of the supernatant (i.e., crude extract) will be the first value for the purification table that will be prepared in Chapter 5. The yeast culture supernatant itself will be the starting point for the purification scheme that was described in Chapter 2.

Materials

Yeast culture broth
Centrifuge
Centrifuge tubes
Biuret reagent (prepared in Chapter 2)
13 × 100 mm test tubes

MORE...

Spectrophotometer
Disposable pipettes and pump/bulb
Micropipettes and tips

Method

1. Centrifuge the remaining culture broth for 5 min. Decant and save the supernatant. Keep the supernatant on ice or refrigerated when not in use.

2. Transfer 1 ml of the supernatant to a tube and add 4 ml of the Biuret reagent. Gently mix the solution and incubate for 20 minutes. Prepare a reagent blank by adding 4 ml of Biuret reagent to 1 ml of distilled water.

3. Zero the spectrophotometer against the blank and measure the absorbance of the solution at 545 nm. Using the standard curve prepared in Chapter 2, calculate the total protein concentration of the supernatant in mg/ml.* If the Biuret reagent used to measure protein concentration in the supernatant is not the same as that used in the experiment from Chapter 2, then a new standard curve will have to be generated.

4. Calculate the specific activity of the yeast supernatant by dividing the α-galactosidase activity/ml of the supernatant by the amount of total protein/ml in the supernatant. The specific activity is reported as enzyme units (U) per mg protein. Record and save this value for construction of the purification table in Chapter 5.

Stability Study of Yeast Supernatant

When working with an unknown enzyme, a series of tests and experiments usually are performed to determine the optimum conditions for

---

*The construction and use of standard curves differ among investigators. With commonly used assays, such as the Biuret protein assay, some researchers will construct one standard curve per batch of reagent. Alternatively, some researchers will construct a standard curve each time the assay is performed. For convenience, we will reuse standard curves.

maintaining enzyme activity (i.e., stability). The yeast extract/peptone broth provides a reasonably stable environment for the α-galactosidase. There are no apparent proteases in the supernatant (extracellular enzyme in the supernatant does not require cell lysis), and the pH/ionic strength of the broth is compatible with maintenance of enzyme activity. α-Galactosidase does not require a reducing environment, and the enzyme can be manipulated at room temperature due to its excellent thermal stability. It is important to note here that enzyme preparations in the real biotechnology world are rarely this accommodating.

Several representative stability experiments are performed in this exercise to demonstrate one possible method for determining storage conditions for the yeast supernatant. Stability experiments will determine the effect of freezing/thawing on the enzyme activity in the yeast supernatant, and glycerol will be added to an aliquot of the yeast supernatant to determine its effect on enzyme stability during the freeze/thaw cycle.

Materials

     Microcentrifuge tubes
     Yeast culture supernatant
     Glycerol
     Disposable pipettes
     Refrigerator/freezer

Method

1. Prepare four 1 ml aliquots of the yeast cell supernatant. Place the remaining supernatant in the refrigerator until the next laboratory session.

2. Aliquot #1 will be stored in a –20°C freezer.

3. Aliquot #2 will be stored at 4°C (in the refrigerator).

4. Add 100 μl of glycerol to Aliquot #3 as a stabilizer. Aliquot #3 is stored at 4°C.

5. Add 100 μl of glycerol to aliquot #4 as a stabilizer. Aliquot #4 is stored at –20°C.

6. If dry ice is available, prepare an acetone/dry ice mixture and freeze aliquots #1 and #4 in the dry ice/acetone slush and then store the quick frozen samples in the freezer. If dry ice is unavailable, just store samples in the freezer. Assay tubes 3 and 4 prior to storage to see if the presence of 10% glycerol has any effect on the α-gal assay.

7. Analyze the four samples for enzyme activity during the next laboratory session. Complete the following table with the data.

Aliquot #	U (before)	U (after)	% U Recovered
1			
2			
3			
4			

STUDY QUESTIONS

1. Based on the results of the enzyme distribution in the yeast culture broth, propose an alternative purification scheme for the isolation and purification of α-galactosidase.

2. Prepare a table of the chemical and biological properties of yeast α-galactosidase, including comments on enzyme stability and optimum conditions for enzyme activity. (Hint: Read both references by Lazo et. al, 1977 and 1978.)

3. How did the activity of the fractionated α-galactosidase compare to the activity associated with the culture broth?

4. Based on the distribution of the α-galactosidase between the supernatant, periplasm, and protoplasts, was the supernatant the best source for the enzyme?

FURTHER READINGS

Perrin D, Dempsey B (1974): *Buffers for pH and Metal Ion Control*. London: Chapman & Hall

Scopes R (1987): *Protein Purification, Principles and Practice*, 2nd Ed. New York: Springer-Verlag

Stoll V, Blanchard J (1990): Buffers: Principles and practice. *Methods in Enzymology* 182:24–38

REFERENCES

Lazo PS, Ochoa AG, Jascón S (1977): α-Galactosidase from *Saccharomyces carlsbergensis*: Cellular localization, and purification of the external enzyme. *Eur J Biochem* 77:375–382

Lazo PS, Ochoa AG, Jascón S (1978): α-Galactosidase (melibiase) from *Saccharomyces carlsbergensis*: Structural and kinetic properties. *Arch Biochem Biophys* 191:316–324

Batch Purification of Proteins

4.1 OVERVIEW

A crude protein extract derived from some type of physical or chemical manipulation of a source will typically contain several types of contaminating biomolecules. These contaminants can include carbohydrates, lipids, nucleic acids, proteins, salts, and other cellular debris. The separation of the protein fraction of the extract from these contaminants is usually referred to as the capture step and is typically performed early in the purification process. The desired protein is then separated from the other captured proteins in subsequent steps using higher resolution purification techniques.

Two popular methods of protein capture are bulk precipitation and batch chromatography. Bulk precipitation methods typically involve salting out the protein fraction by the addition of chaotropic salts, such as ammonium sulfate. The protein fraction can also be precipitated by the addition of certain organic solvents or by long chain synthetic polymers. Batch chromatography involves the binding of the protein fraction to some type of chromatographic gel, followed by filtration to remove contaminants, and, finally, elution and collection of the captured proteins.

The crude α-galactosidase extract has now been prepared, and the conditions for maintaining activity have been investigated. α-Galactosidase will now be isolated, or captured, from the crude extract prior to its purification and analysis. This capture will effectively separate gross contaminants such as cellular debris, nucleic acids, and lipids from α-galactosidase. The experimental section in this chapter will consist of the capture of the α-galactosidase component from the crude extract by batch ion exchange chromatography.

4.2　BACKGROUND

The biological and chemical properties of proteins are exploited in the process of their isolation and purification. The principal chemical properties of proteins that are utilized are solubility in salt solution, charge, size, hydrophobicity, and biological affinity.

The initial capture is an early step in the overall purification scheme, and the degree of purity achieved in the capture can be compromised for the sake of speed and capacity. The goal of this capture process is to isolate the protein(s) from the crude extract with good recovery in a minimal amount of time. The capture procedure generally produces a crude protein preparation (with the exception of some affinity techniques) free of gross contaminants but still requiring further purification.

Bulk Precipitation Techniques

One of the more popular methods of protein capture is bulk precipitation from the crude extract. The surface of a globular protein generally consists of water soluble hydrophilic amino acid residues. Following the convention of "like dissolves like," most globular proteins are soluble in water or salt solutions due to the predominance of these hydrophilic amino acid residues on the protein surface. The solubility properties of proteins in salt solutions can be used to fractionate, or capture the protein from the crude extract via precipitation.

Precipitation is typically accomplished by increasing the ionic strength (i.e., salt concentration) of the crude protein extract by adding ammonium (or sodium) sulfate. By increasing the amount of salt in the protein extract, water is increasingly consumed by solvation of the ammonium and sulfate ions. This leaves less water available for solubilization of the protein. At some point, a limit is reached, and the protein is precipitated, or salted out. Although the salt that is typically used in this procedure is ammonium sulfate, other reagents such as polyethylene glycol and organic solvents (e.g., acetone and ethanol) have also been used for the bulk precipitation of proteins.

In this way, the protein fraction of a crude extract can be captured, or isolated from the other contaminating biomolecules such as lipids and nucleic acids. The method can sometimes be used to fractionate the protein of interest from the other protein contaminants in the extract if the salt is added slowly in discrete steps. The protein precipitate can be recovered after each ammonium sulfate addition, or cut, by centrifugation.

Batch Protein Capture

Another popular method for bulk protein capture from the crude extract involves the use of the chromatographic technique of batch ion exchange chromatography. Other batch chromatography techniques can also be employed, but ion exchange is the most widely used.

As previously mentioned, globular proteins typically have the hydrophilic amino acid residues exposed on their surface which facilitates solubility in aqueous solutions. Some of these residues contain acidic and basic side chains which give the protein an overall net charge. The acidic amino acids are glutamic and aspartic acids, and the basic amino acids are lysine, arginine, and, to a lesser extent, histidine. A typical protein that contains both acidic and basic amino acids will have an overall positive charge under acidic buffer conditions (at low pH) and an overall negative charge under more basic buffer conditions (at high pH). The transition between these two extremes is referred to as the protein isoelectric point, or pI, and it is defined as the pH where the protein assumes a net charge of zero. The pI value is an important parameter and should be determined prior to setting up any rational protein purification scheme. The pI values of some familiar proteins are shown in Table 4.1.

Batch ion exchange chromatography takes advantage of the ionic properties of proteins to capture the protein from the other components in the crude extract. An ion exchange gel (ion exchanger), the material

Table 4.1 Isoelectric Point of Some Typical Proteins

Protein	Isoelectric Point
Pepsinogen	2.80
Amyloglucosidase	3.50
Glucose oxidase	4.15
Soybean trypsin inhibitor	4.55
β-Lactoglobulin A	5.20
Bovine carbonic anhydrase B	5.85
Human carbonic anhydrase B	6.55
Horse myoglobin (isozyme)	6.85
Horse myoglobin (isozyme)	7.35
Trypsinogen	9.30
Cytochrome C	10.25

used to capture the protein, consists of a cationic or anionic ligand chemi-cally grafted to a cross-linked polymer, usually in beaded form. The ligand is typically a small molecule covalently attached to the gel that is responsible for the ionic properties (i.e., the ligand is the component of the exchanger that gives the gel its charge). The beaded ion exchange gels are used almost exclusively in column chromatography because of their excellent flow properties; however, in the batch mode, the gel is simply added to the crude extract in a beaker or other suitable vessel. The batch method, therefore, does not require beaded ion exchange gels since the flow properties of the gel are not important. Column chroma-tography, where beaded gels are used almost exclusively, will be covered in greater detail in Chapter 5.

The polymers that are the backbone of the ion exchange gels are generally based on dextran or cellulose, but exchangers grafted to poly-styrene are also used, especially for the separation of small molecules such as amino acids. Ion exchange gels based on dextran and cellulose are commercially available, and each is supplied as a solid that, when added to aqueous buffer, does not dissolve but swells to several times its dry volume.

Agarose based ion exchange gels are also commercially available and are supplied preswollen. The preswollen agarose gels have much better flow properties and are more convenient to work with but are also more expensive. Beaded agarose based gels are rarely used in batch ion exchange chromatography.

Several types of charged ligands are commercially available. Anionic ligands such as carboxymethyl (CM) or sulfonic acid (S), bind proteins that are positively charged, and the process is called cation exchange chromatography (Figure 4.1).

Cationic ligands, such as diethylaminoethyl (DEAE) or quaternary ammonium (QA), bind proteins that are negatively charged and is referred to as anion exchange chromatography (Figure 4.2).

The charge characteristics of the protein that is to be purified are determined experimentally. This involves measuring the net charge of a pro-

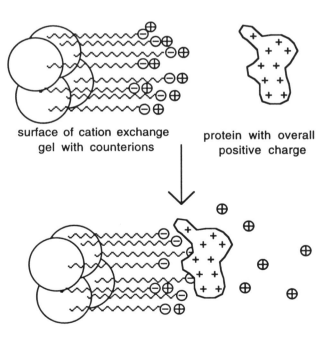

surface of cation exchange
gel with counterions

protein with overall
positive charge

protein binds to charged ligand and displaces counterions

Figure 4.1 Schematic of cation exchange mechanism.

tein at a given pH by isoelectric focusing or by protein capture assays (see below). The ion exchange gel is then added to the protein in a solution of low ionic strength binding buffer, and the protein binds to the gel. The actual exchange occurs as the protein binds to the gel and displaces the counter ion originally bound to the gel.

The ion exchange gel, with bound protein, is washed with more binding buffer and is then treated with elution buffer usually composed of the same buffer solution but containing a high concentration of a salt, such as sodium chloride. The increased ionic strength of the elution buffer releases, or elutes, the protein from the gel (Figure 4.3).

In this way, the protein components of the crude extract are separated from the other contaminants. Separation of the desired protein from the other contaminating proteins in the extract does not usually occur under batch conditions since the purpose of the step is to capture the protein fraction from the crude extract rapidly and with high yield. This

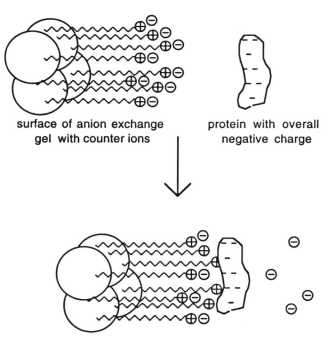

surface of anion exchange
gel with counter ions

protein with overall
negative charge

protein binds to charged ligand and displaces counter ions

Figure 4.2 Schematic of anion exchange mechanism.

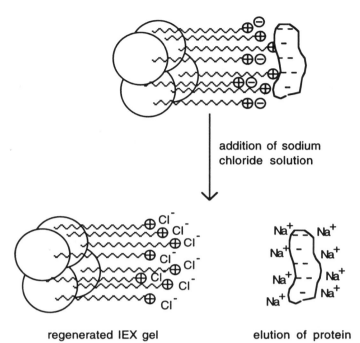

Figure 4.3 Elution of protein from an ion exchange gel.

regenerated IEX gel

elution of protein

is especially true if the protein is unstable. The binding and elution mechanisms for ion exchange are essentially the same in both batch and column methods.

Gel Binding Capacity

There are many factors to consider when choosing a gel for protein capture by batch ion exchange chromatography. Properties of the gel such as particle size, monodispersity (percent variation of the gel bead size) and shape (i.e., beaded or not) are not important considerations in the batch mode, but they are important characteristics in column chromatography. In Chapter 5, when column ion exchange chromatography is discussed, these properties of the gel will be considered in more detail.

The ion exchange gel should have a sufficiently large available capacity to make it practical under the conditions and scale of the purification. The ionic capacity of a gel refers to the density of the charged ligands and is usually reported as mole equivalent of charged ligand per gram of dry gel or per ml of swollen gel. The available capacity, however, refers to the maximum amount of protein that will bind to the gel under defined experimental conditions and is generally reported as mg protein/ml gel. The available capacity of the gel is dependent on three criteria: (1) ionic capacity, or charge density; (2) pore size of the gel; and (3) the nature of the gel ligand (i.e., strong versus weak exchanger).

The ionic capacity is defined as the concentration, or density, of charged ligand molecules bound to the ion exchange gel. The greater the density of the ion exchange ligand on the gel, the more protein the gel will bind. The correlation of ligand density to protein binding, however, is not one to one because the protein is large and usually interacts with more than one ligand on the gel at a time. Even if protein to ligand binding is one to one, steric interactions between proteins can limit binding capacity. The actual binding capacity of the protein to the gel, i.e., available capacity, can only be accurately determined experimentally. Obviously, this is not a convenient situation, and manufacturers generally supply the average, or approximate available capacity of the gel. The available capacity of the gel is determined by binding studies using standard proteins such as bovine serum albumin or gamma globulin. Therefore, the available capacity reported is only an approximate value.

Pore size is another factor which influences gel capacity. The number of ion exchange ligands on the surface of the gel is only a small percentage of the total ion exchange capacity. The protein must be small enough to diffuse into the gel particle itself in order to bind to the charged ligands in the interior of the gel bead. In this way, the protein will utilize all of the ion exchange capacity of the gel. Commercially available ion exchange gels cover a range of pore sizes, and selection of the correct gel is crucial for insuring good recovery.

The third factor important in capacity considerations is whether the ion exchange gel contains a weak or a strong ion exchange ligand. Strong ion exchange gels contain ligands, such as quaternary ammonium groups, that maintain their charge independent of buffer pH. Weak exchangers, such as diethylaminoethyl (DEAE), contain an ionizable proton and therefore, the charge on the ligand is pH dependent. The essential difference between the two is that the strong ion exchangers maintain their ionic capacity over the entire working range of pH whereas the ionic capacity of the weak ion exchangers varies with the pH of the buffer. The available capacity of the strong exchangers, therefore, does not vary with pH, unlike the available capacity of the weak exchanger, which does.

4.3 EXPERIMENTAL DESIGN AND PROCEDURES

In this chapter, the technique of batch ion exchange chromatography will be used to capture α-galactosidase from the yeast culture supernatant. This capture is being done as an alternative to precipitation by the addition of ammonium sulfate. Ammonium sulfate techniques are more cumbersome and require resolubilization of the enzyme from the high salt precipitate, so we have decided to use the more efficient batch ion exchange technique.

Determination of Binding pH

The ion exchange gel that we will use for the capture of the α-galactosidase from the yeast culture supernatant is DEAE Sephadex® A-50 produced by Pharmacia Biotech. This ion exchange gel is composed of spherical carbohydrate particles which have relatively large pores.

The experimental conditions for the actual binding of the enzyme must be empirically determined. Ideally, the initial binding of the enzyme to the gel takes place under conditions of low ionic strength at a pH where: (1) the enzyme is stable; (2) the enzyme binds to the gel; and (3) the ion exchange gel has a sufficient available capacity.

Our protein, α-galactosidase, exhibits maximum enzyme activity at pH 4.5 and is stable over a broad range of pH. The DEAE Sephadex gel has a working pH range of approximately 2 to 9. This means that within this range the gel has good binding capacity for enzyme capture. Below pH 2.0, the ligand is still charged, but the Sephadex gel matrix itself becomes unstable and breaks down. Above pH 9.0, the DEAE group begins to lose its charge, and, therefore, the enzyme binding capacity is diminished.

The parameter that must be determined is the pH at which α-galactosidase actually binds to the gel. Remember that proteins possess a net charge of zero at the isoelectric point, that is, the pI. Proteins are positively charged in buffers with a pH below the pI and negatively charged in buffers at pH above the pI. The DEAE ion exchange gel is an anion exchange gel; therefore, the enzyme must possess an overall negative charge (pH of the buffer > pI) in order to bind to the gel.

The pI of the enzyme can be determined in several ways, including isoelectric focusing gel electrophoresis. This technique involves the migration of proteins through a polyacrylamide gel, in which the gel behaves as a pH gradient. The proteins will only migrate if they are charged, and when the proteins move into a pH at which they become neutral, they stop migrating. The pH at which migration of the protein ceases is that protein's pI.

A simpler, qualitative test tube method can also be used to determine the pH where α-galactosidase binds to the ion exchange gel. A set of test tubes is prepared with each tube containing a small amount of the DEAE Sephadex A-50 ion exchange gel. Using a series of buffers, the gel in each tube is equilibrated to a different pH, each tube differing by about 0.5–1.0 pH units. A series of the crude culture supernatant is also adjusted to the same pH. During this procedure, the salt concentration (ionic strength) of each tube is kept low (25 mM) in order to insure binding (the enzyme will not bind to the gel if the ionic strength of the buffer is too high). The enzyme and ion exchange gel series are mixed accordingly and incubated for five minutes. The gel is allowed to settle, and then the supernatant from each tube is analyzed using the α-galactosidase assay. A positive binding result is indicated when the supernatant contains little or no enzyme activity. The lack of activity in the supernatant indicates that binding to the ion exchange gel occurred at that pH.

A test run using myoglobin will first be performed to demonstrate the technique. The myoglobin binding can be detected by visual inspection of the supernatant and the gel. Myoglobin has a brownish red color and binding is indicated by the gel turning from white to the brownish color of the myoglobin solution.

A similar set of experiments is normally done to determine the ionic strength at which binding of the enzyme will occur. The same protocol is followed except that the pH is kept constant, and the ionic strength of each tube is varied. Time constraints do not permit the execution of this experiment so the assumption is made that the ionic strength of the buffers must be below 50 mM for enzyme binding to occur.

Materials

        13×100 mm test tubes
        DEAE Sephadex A-50
        0.1 M citrate, pH 3.0 and 4.0 buffers
        0.1 M Tris-HCl, pH 7.0, 8.0 and 9.0 buffer
        0.1 M bis-Tris, pH 5.0 and 6.0 buffer
        Myoglobin solution, 1 mg/ml in water
        Vortex mixer
        Yeast culture supernatant
        Microcentrifuge tubes
        Micropipettes and tips
        0.5 M sodium acetate, pH 4.5
        0.1 M p-nitrophenyl-α-D-galactoside, in water
        0.1 M sodium carbonate buffer
        Spectrophotometer

Method

1. Set up eight 1.5 ml microcentrifuge tubes for pH 3, 4, 5, 6, 7, 8 and 9. The eighth tube will be a blank.

2. Add approximately 25 mg of DEAE Sephadex A-50 gel to each tube (25 mg is approximately enough gel to fit on the tip of a micro spatula). Alternatively, if the gel is already swollen, add 100 μl of gel slurry to each tube.

3. Wash the gel in each tube three times with the appropriate pH buffer diluted to 0.025 M. The gel is washed by adding 1 ml of the buffer, lightly vortexing, centrifuging for 1 minute at 3000 rpm, and decanting.

4. In separate tubes, mix 750 μl of the myoglobin solution with 250 μl of each of the pH buffers. Add those myoglobin standards to the corresponding microfuge tube with the ion exchange gel, mix, and allow it to incubate for 5 minutes. Spin down the gel and visually determine at what pH binding occurs. Myoglobin is brownish red, and the gel becomes discolored brownish red as binding occurs.

5. For measuring the binding of α-galactosidase, repeat steps 1–3. In separate tubes, mix 250 µl of supernatant, 250 µl of each of the seven buffers, and 500 µl of water. Add each of the buffered enzyme solution to the tube containing its corresponding buffered gel. Incubate the mixture for 10 minutes. Spin down the tubes and analyze the supernatant using the α-galactosidase enzyme assay. The lack of enzyme activity in the supernatant indicates a positive result meaning that the α-galactosidase binds to the gel at that pH.

Capture of α-Galactosidase by Batch Ion Exchange

Once suitable buffer pH conditions for the binding of the α-galactosidase from the yeast culture supernatant have been determined, the capture can be performed. The remaining yeast broth supernatant ($\approx$ 45 ml) will be diluted with distilled water, and the pH adjusted to the pH that allows for efficient binding of the protein to the ion exchange gel. The purpose of the dilution with distilled water is to insure that the ionic strength of the supernatant is maintained below 50 mM salt (ionic strength). The DEAE Sephadex A-50 gel slurry is added to the supernatant and incubated for one hour. The gel/enzyme slurry is filtered, collected, and then treated with an elution buffer (containing sodium chloride to release the bound proteins from the gel). This elution step is repeated two more times. The α-galactosidase activity of each fraction will be determined by enzyme assay. The specific activity of this fraction will be calculated after the total protein content is determined by the Biuret protein assay.

The degree of purification is determined by a comparison of the specific activities of the crude supernatant and captured fractions. The results will be recorded in a α-galactosidase purification table, and the yield of the capture step will be determined by a comparison of the total activity units recovered. A sample purification table is shown below in Table 4.2.

Materials

Yeast culture supernatent

1.0 M Tris base (pH not adjusted)

MORE...

Table 4.2 Sample Purification Table

Step	Volume (ml)	Total Activity	Total Protein	Specific Activity	Yield (%)	Purifica-tion
Supernatant	500	52.2	515	0.102	100	1
DEAE-Sephadex Batch	30	36.4	34	1.07	69	10
DEAE-Sephadex Chromatography	6	25.5	16.7	1.53	49	15
Gel Filtration Chromatography	8	21.2	11.3	1.88	40	18

DEAE Sephadex A-50 gel slurry/preswell at least 24 hr in advance*

Whatman filter paper

pH Paper or pH Meter

30 ml sintered glass funnel, medium porosity

25 mM Tris-HCl, pH 7.5, 0.5 M NaCl

13 × 100 mm test tubes

0.5 M sodium acetate, pH 4.5

Constant temperature block or water bath

0.1 M p-nitrophenyl-α-D-galactoside, in water

0.1 M sodium carbonate buffer

Spectrophotometer

Method

1. Sephadex based ion exchangers must be swollen (i.e., equilibrated) in buffer before they can be used. The DEAE Sephadex A-50 gel slurry is prepared by mixing three (3) grams of the powdered gel in about 250 ml of 25 mM Tris-HCl, pH 7.5, and allowing it to swell overnight at room temperature. The gel sediments by gravity to a volume of

*Each person/group will require no more than 20 ml of swollen DEAE Sephadex A50 for the experiments in both Chapters 4 and 5. Proportionally, 3 grams of dry gel will swell in 25 mM Tris-HCl, pH 7.5 buffer, to yield a final volume of 100 ml, following the protocol described below. Unused swollen Sephadex can be stored in 0.002% chlorohexidine for prolonged periods.

about 90 milliliters. Decant the clear buffer above the gel bed so that the final volume of the slurry is about 100 ml (total volume).

2. Dilute the yeast culture supernatant (≈ 45 ml) to 200 ml with distilled water in a 500 ml flask. Adjust the pH of the diluted supernatant to 7.5 using 1 M Tris base. This step requires approximately 0.5 to 1.0 ml of the Tris base to reach pH 7.5.

3. Add 12 ml of DEAE Sephadex A-50 gel slurry to the diluted supernatant. Shake and allow the gel to incubate for 1 hour with frequent swirling of the mixture.

4. After one hour, the enzyme/gel slurry is filtered, and the gel is collected. The gel can be filtered and collected using a sintered glass funnel (medium porosity) connected to a water aspirator. (Be careful not to allow the gel to dry.) Then the gel is washed with 25 mM Tris-HCl, pH 7.5, to insure removal of any unbound protein. Alternatively, the gel slurry can simply be filtered by gravity using Whatman filter paper. When using the filter paper technique, the washing with 25 mM Tris-HCl, pH 7.5, is done carefully in order that the gel concentrates from the sides of the filter paper to the center of the filter paper cone.

5. The gel is then washed and eluted with approximately 30 ml of elution buffer consisting of 25 mM Tris-HCl, pH 7.5, containing 0.5 M sodium chloride. The gel is washed three times each with 10 ml of the elution buffer (for a total of 30 ml). The high salt washings are collected (but not combined), and the three samples are labelled Gal-1 (a–c). Each sample is then assayed for α-galactosidase activity.

6. Assay the Gal-1 (a–c) fractions for total protein content by the Biuret method as described in Chapter 3. The Biuret reagent will react slightly with the Tris buffer that was used to elute the α-galactosidase from the gel. A suitable negative control is to perform a total protein assay on the 25 mM Tris, pH 7.5, 0.5 M NaCl. The resulting absorbance can be subtracted from the absorbance obtained from the enzyme fractions. In this way the blue color resulting from the presence of Tris buffer is factored out. Alternatively, the elution buffer/ Biuret reagent can be used to zero the spectrophotometer.

Construct a purification table similar to Table 4.2. Record the data in the purification table. Calculate the purification factor for this step and record it in the table.

STUDY QUESTIONS

1. Describe a cation exchange purification protocol for a protein that has a suspected pI of approximately 7.0. Include in your description test tube assay conditions, gel buffer requirements, etc.

2. Why is it important to rapidly capture the protein immediately after it is isolated from its biological source?

3. What would be the effect on the batch purification of α-galactosidase if:

   (a) the ionic capacity of the gel were increased?

   (b) the pore size of the gel was decreased?

   (c) the gel was changed to a strong anion exchanger?

   (d) the gel was changed to a beaded cation exchange gel?

4. Why was the yeast supernatent diluted to 200 ml before adding the gel slurry? What is an alternative to dilution of the sample? (Hint: See Background section of Chapter 5.)

FURTHER READINGS

Scopes R (1987): *Protein Purification, Principles and Practice*, 2nd Ed. New York: Springer-Verlag

REFERENCES

Lazo PS, Ochoa AG, and Jascón S (1977): α-Galactosidase from *Saccharomyces carlsbergensis*: Cellular localization, and purification of the external enzyme. *Eur J Biochem* 77:375–382

5

Protein Purification by Column Chromatography

5.1 OVERVIEW

Following the removal of the gross contaminants from the crude extract, the remaining protein components must be resolved. The most popular and ubiquitous technique for this separation is column chromatography. Similar to batch purification, column chromatography utilizes chemical and biological properties of the protein for its purification, but produces greater resolution.

The background section of this chapter will cover some of the routine column chromatographic techniques used in protein purification. This discussion will elaborate on the material presented previously in Chapter 4 concerning chromatographic theory and methods.

In the last experiment, α-galactosidase was captured from the crude extract and assessed for its purity. The increase in specific activity of the enzyme over the original supernatant demonstrates an initial purification. The enzyme must now be further purified to apparent homogeneity prior to its analysis. The experiment for this chapter will be to purify α-galactosidase by ion exchange column chromatography using a step gradient of sodium chloride to elute the enzyme. The α-galactosidase will be assessed for purity by determining its specific activity.

5.2   BACKGROUND

Column chromatography is by far the most commonly used technique for protein purification. Similar to batch techniques, purification is accomplished by exploiting the physical and chemical properties of proteins such as size, charge, hydrophobicity and also, biological affinity. Batch chromatographic techniques differ from the column techniques, however, in that batch techniques sacrifice resolution for speed, convenience, and capacity. Column chromatographic methods are experimentally more complex than the batch methods, but they provide the necessary resolution for the final purification of the protein to apparent homogeneity.

Components of a Chromatography System

A simplified column chromatography system has a buffer reservoir(s), a fluid delivery system (pump), a column with chromatographic gel, a means of detecting molecules eluting from the column, and a collecting system (Figure 5.1).

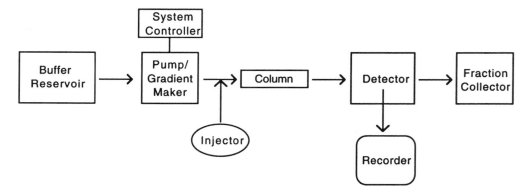

Figure 5.1   Schematic of a simple chromatographic system.

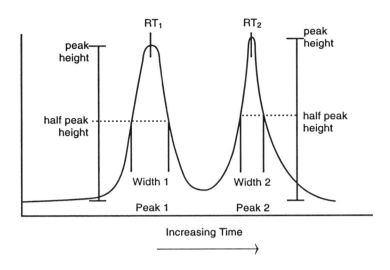

Figure 5.2 Sample chromatogram showing resolution of peaks.

A typical column chromatography experiment consists of applying a sample to a column filled with a specific chromatographic gel (matrix), elution of the column with a suitable buffer, and detection of the eluting components followed by collection of the desired purified product. The level of instrument sophistication will vary depending on the level of performance desired, but the above sequence of events is relatively standard.

Chromatographic Resolution

The goal of all column chromatography is to resolve the individual components in a crude solution by application of a sample of the solution to the column and by elution with an appropriate buffer. Resolution is defined as two times the distance between two peaks eluting from the column divided by the sum of the peak widths at half height (Equation 5.1). A sample chromatogram illustrates the principle of chromatographic resolution (Figure 5.2).

Equation 5.1   Formula for resolution

$$\text{Resolution} = R_s = 2 \times \frac{R_{T\,\text{Peak 1}} - R_{T\,\text{Peak 2}}}{W_{H\,\text{Peak 1}} + W_{H\,\text{Peak 2}}}$$

R_T = retention time of peak
W_H = peak width at half height

Resolution can also be defined as the selectivity of the chromato-graphic method divided by the efficiency of the column. Selectivity is essentially a chemical process (i.e., it depends on the nature of the gel packed in the column), and it is defined by the distance between two chromatographic peaks regardless of the peak widths. The selectivity determines the retention of the eluting component and is defined in units of time or volume.

The efficiency is a reflection of the physical components of the column and system and is defined by the width of the individual peaks regard-less of their separation. The narrower the peak width (greater efficiency), the less the sample was diluted as it traversed the column. This narrow peak represents low dilution and is the result of efficient kinetics (i.e., rapid diffusion) during the actual separation mechanism.

In a typical protein purification, the important parameter to maxi-mize is the resolution of the protein peak(s) of interest from the other contaminating proteins in the sample. Resolution can be maximized by one of three methods. The first method is to increase the distance between two eluting peaks by changing the selectivity of the system. An example of this would be to change from a Q type anion exchange gel to an S type cation exchange gel (see Figure 5.3A). The second method increases the efficiency of the separation by changing a physical parameter such as the particle size of the gel bead. Small gel beads result in better kinetics and less dilution of the protein peak as it elutes from the column. The result is an improvement in resolution due to more narrow peak widths, even though the peak retention may remain essentially unchanged (see Figure 5.3B). The smaller the particle size, however, the greater resis-tance there is to buffer flow so the pressure within the column will increase. The third method is a combination of A and B in Figure 5.3.

The factors that determine the efficiency of a chromtographic experi-ment are virtually the same for all types of column chromatography.

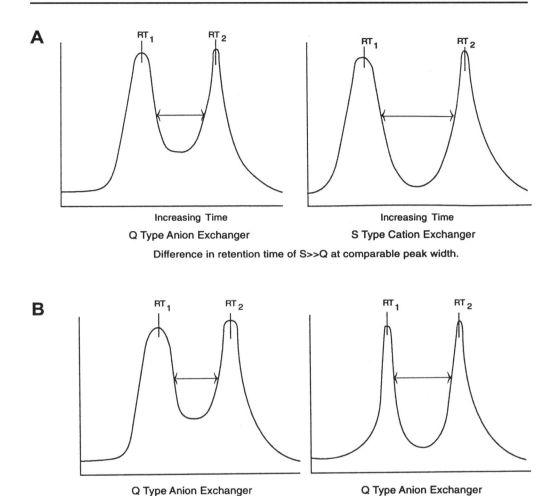

Figure 5.3 (A) Example of changing selectivity of gel. (B) Effect of particle size on resolution.

The selectivity, however, is dependent on the actual chemical process responsible for the separation, i.e., the retention mechanism. The retention mechanism exploits typical protein properties of size, charge, affinity, etc., to effect a separation.

Gel Filtration Chromatography

Gel filtration (GF) chromatography separates globular proteins by size (i.e., molecular weight) and consists of a column packed with porous polymeric beads. These beads are usually composed of cross-linked polysaccharides (some GF media use synthetic polymers such as polyacrylamide) in which the pore sizes within the gel beads are controlled.

A sample is applied to the top of a column filled with gel, and the protein is eluted by passing a defined volume of a suitable buffer through the gel matrix. The larger proteins cannot physically enter the small pores of the beads, and, therefore, spend less time in the gel bead and more time in the eluting buffer. The eluting buffer is sometimes referred to as the mobile phase, and the gel itself is called the stationary phase. Each bead is similar to a three dimensional maze that retards the elution of the smaller molecules. As such, the larger molecules elute from the column first, followed by the smaller molecules, and the result is the fractionation of the protein mixture by size. The size of a globular protein is related to its molecular weight so gel filtration effectively separates proteins by the differences in their molecular weight (Figure 5.4). In this way, gel filtration chromatography can be used to estimate the molecular weight of a protein. The retention times (or volumes) of a series of globular proteins with known molecular weights are determined. The data are used to plot log molecular weight versus retention time to generate a calibration curve that can be used to estimate the molecular weight of an unknown protein. The linear relationship breaks down with nonglobular proteins and DNA since both are nonspherical in shape.

The resolution in gel filtration is dependent on column length, the flow rate of elution, pore size of the gel, and the particle size of the gel bead. In general, the efficiency, and therefore the resolution, of any chromatographic analysis improves as the size of the gel bead decreases. The pore size of the gel determines the molecular weight range over which separation will occur. The column length is important in gel filtration, and resolution generally increases with increasing column length. However, increasing column length eventually degrades resolution due to diffusion of the sample as it transverses the gel filtration column. In addition to good recovery and mild conditions, gel filtration usually gives good resolution and is relatively easy to perform.

Due to the dilution of the proteins, gel filtration is limited by the volume of sample that can be applied to the column. The sample volume

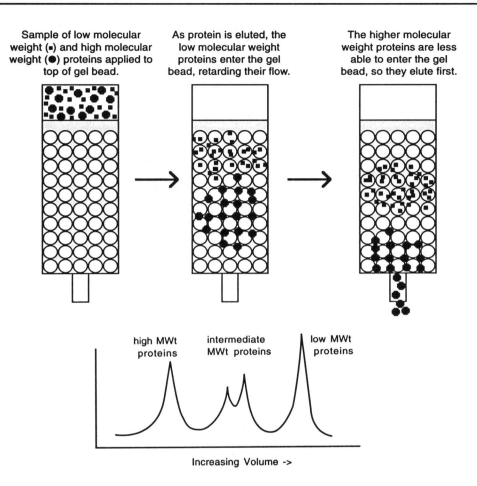

Figure 5.4 Mechanism of gel filtration.

should be between 2–5% of the total column volume. Larger sample volumes lower the resolution of the technique. The column volume is defined as the total volume occupied within the column by the gel slurry, and it can be measured empirically by filling the empty column with water and measuring its volume. Alternatively, the total column volume is equal to the volume required to elute buffer salts from the column since the small salt molecules permeate the entire gel bead.

In addition to fractionation of proteins by size, gel filtration can be used as a sample preparation technique for buffer exchange or desalting. The principle is the same because the molecular weight difference between the protein sample and the buffer salts can be used to completely separate the two components. As we will see in the experimental section, this is an extremely important technique and is routinely used in the laboratory to prepare protein samples for chromatography and electrophoresis. The sample volume for desalting can be as high as 25% of the column volume. The greater sample volume is possible because the resolution requirements are not as stringent as for normal gel filtration chromatography. The decrease in resolution is a result of the decrease in efficiency due to the larger sample volume. The compromise in resolution is offset by the increase in sample capacity for the protein desalting.

A major disadvantage of gel filtration is that the sample elutes from the column at a lower concentration due to dilution of the sample in the gel during the separation. The technique, however, is performed under extremely mild conditions with excellent recovery of protein mass and biological activity.

Ion Exchange Chromatography

Ion exchange chromatography (IEX) is both a high resolution and high capacity purification technique. The separation is based on the overall charge of the protein, which is determined by the number of acidic and basic residues in the protein, and the pH and ionic strength of the eluting buffer. Ion exchange gels for column chromatography are commercially available for either cation or anion chromatography. The terms weak and strong are used to classify both anion and cation exchanger. As mentioned previously, the strong exchangers, such as sulfonic acids and quaternary ammonium salts, retain their charge over a wide pH range. The result is that the ionic capacity of a strong ion exchanger gel is independent of buffer pH since, as long as the exchanger is charged, it can bind protein. The weak exchangers, such as carboxymethyl or diethylaminoethyl, are charged over a limited pH range so that the ionic capacity of these gels does vary with buffer pH. At a pH at which the exchanger is minimally charged, proteins may not bind well, and as that pH is approached, the available capacity of the exchanger steadily decreases due to a reduction of the number of charged ligands on the gel. Particular buffers may also interact with the gel and affect its capacity. To prevent buffer interactions with the gel, it is advisable to use anionic buffers (e.g., phosphate) for cation exchange and cationic buffers (e.g., Tris) for anion exchange.

In a typical ion exchange experiment, the protein sample is adjusted to an appropriate pH and ionic strength so that it will bind to the gel. The sample is then applied to the top of the column where it immediately binds to the gel, and the result is a concentrating effect. When performing column chromatography, the available capacity of a gel is dependent on the flow rate used to load the sample; thus the choice of flow rate is very important. The maximum amount of protein that will bind is referred to as the dynamic capacity and is defined as the available capacity at a specific flow rate and a specific column geometry and packing. Due to the concentrating effect, the starting volume of the sample in ion exchange, unlike in gel filtration, is not a critical factor for resolution. Next, the protein is eluted from the column, usually by changing the pH or increasing the salt concentration of the buffer (Figure 5.5). This can be done at a steadily increasing rate (i.e., a linear gradient) or in sharp increments (i.e., a step gradient).

Linear gradients usually result in better resolution than step gradients but require more sophisticated instrumentation. Linear gradients are formed by combining a buffer with low ionic strength with a buffer of high ionic strength in specialized gradient makers. A schematic of the column ion exchange chromatography mechanism is presented in Figure 5.5.

The parameters affecting resolution in ion exchange chromatography are gel selectivity and particle size, flow rate, gradient slope, and choice of pH. The column length in gradient ion exchange chromatography is not a critical factor, and, in fact, the shorter columns generally give better results due to reduced dilution of the sample upon elution. Resolution increases with decreasing particle size; however, the high back pressure associated with small particle size requires expensive columns and instrumentation. Gradient slope and flow rate must be optimized empirically although shallow gradients and average flow rates (e.g., 1 ml/min) generally yield good results. Ion exchange is an extremely popular technique because it combines high available capacity and high resolution under chromatographic conditions that are mild and do not denature the protein.

Hydrophobic Interaction Chromatography

Hydrophobic interaction chromatography (HIC) is another technique that combines high capacity and high resolution under non-denaturing conditions. The gels that are commercially available generally contain either a phenyl ligand or a short chain, branched hydrocarbon ligand. The ligand is covalently attached to a cross-linked gel that is usually polysaccharide based. HIC is similar to IEX in that it can utilize either a

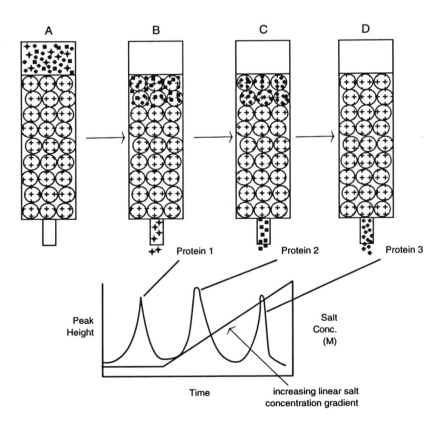

Figure 5.5 Column ion exchange chromatography. (A) The protein solution is loaded onto an equilibrated IEX gel. (B) Protein 1 washes through while Proteins 2 and 3 bind. (C) Increasing concentrations of salt displace Protein 2 which elutes. (D) Further increasing salt concentration displaces Protein 3.

linear or a step gradient for elution. The protein binds to the gel bound ligand by hydrophobic interaction (Figure 5.6).

A typical HIC experiment begins by equilibrating the protein sample in an appropriate buffer and loading the sample onto the column where it immediately binds to the gel. Similarly to what happens in ion exchange, the binding of the protein to the gel results in a concentrating effect so that dilute samples can be applied, and sample volume is essentially unimportant. In contrast to ion exchange, the loading and eluting buffer contains high concentrations of ammonium sulfate or sodium sulfate salts, usually at a concentration between 1.0–2.0 M. The purpose of the salt is to raise the ionic strength of the protein solution so that the hydrophobic binding of the protein to the gel is induced. Additionally, the sulfate salts act as chaotropes, that is, they interfere with protein solvation and enhance binding. The hydrophobic effect here is similar to the salting out of the protein using ammonium sulfate (Chapter 4), and the binding is considered to involve only those hydrophobic residues located in the so-called patches on the surface of the globular protein. The hydrophobic groups buried in the interior of the protein are not believed to be directly involved in binding so no denaturing of the protein occurs. The protein is typically eluted from the column using a linear gradient of **decreasing salt** concentration.

In hydrophobic interaction the protein sample goes from high salt to low salt environment. This is the exact opposite of ion exchange in which elution is from low salt to high salt. This feature can allow for HIC and IEX (or vice versa) purification in series without sample pretreatment (i.e., desalting). Detergents are sometimes added to the HIC elution buffer to help maintain solubility of the protein.

Resolution in hydrophobic interaction chromatography is dependent on the nature of the hydrophobic gel ligand, particle size, buffer pH, gradient slope, and flow rate. The column length in HIC is not critical when using gradient elution. When detergent is required, the nature and concentration of the detergent can affect resolution.

Affinity Chromatography

Affinity chromatography is by far the most powerful chromatographic technique with respect to resolution and specificity. As mentioned previously, proteins are defined by their chemical and biological properties.

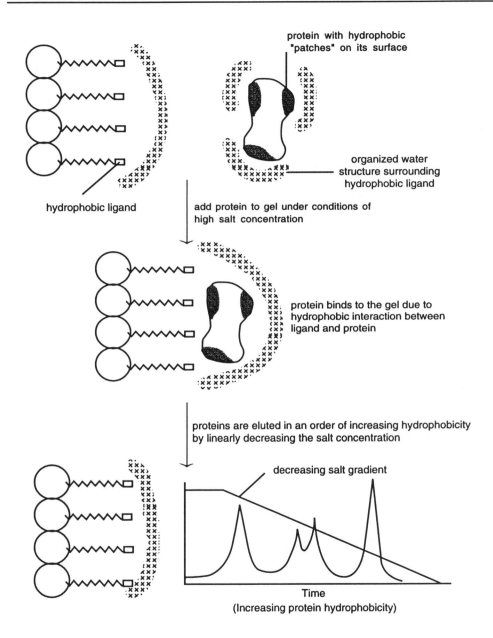

Figure 5.6   Mechanism for protein binding in HIC.

Affinity techniques exploit the biological properties of the protein in that the protein of interest specifically binds to an immobilized ligand via some specific biological affinity. All the other proteins in the mixture remain in solution and pass through the column.

The binding is based on the exploitation of some specific biochemical property of the desired protein, for example, the specific binding of an antibody to a gel bound antigen. Affinity techniques differ from chemical techniques in that they exploit individual biochemical properties specific to a particular protein. The chemical techniques, such as ion exchange or HIC, exploit global properties of the protein (e.g., charge, size, hydrophobicity) that are shared to some degree by all other proteins in the mixture.

The affinity techniques are extremely powerful but limited in general application. Whereas an ion exchange gel can be used for virtually all types of proteins or protein preparations, affinity gels usually work only for a single type of protein. The attractive feature of affinity chromatography is that the desired protein can be purified to biochemical homogeneity in virtually a single step.

Exceptions to this broad generality are several. For example, Protein A and Protein G are derived from *Staphylococcus aureus* and *Streptomyces griseus*, respectively, both of which bind to the heavy chain constant region of IgG. Protein G or Protein A ligands attached to a gel will purify most immunoglobulins that belong to the IgG class (with the exception of some subclasses) regardless of their antigenic specificity. Protein A and Protein G are examples of affinity techniques in which the specificity is not limited to one particular protein but is specific for an entire class of proteins.

Another example of a group affinity technique is chromatography using immobilized lectins. Lectins are carbohydrate binding proteins (e.g., wheat germ agglutinin which binds N-acetylglucosamine residues and oligomers) that are used to purify carbohydrates and glycoproteins. The binding occurs between the carbohydrate group of the desired glycoprotein and the immobilized lectin. The binding, therefore, is not specific to a particular protein but to the entire class of proteins that contain the common structural feature of a carbohydrate group.

Column affinity chromatography differs from batch chromatography in that the application of a gradient can sometimes separate subclasses of proteins. For example, a Protein A affinity gel can be used to separate subclasses of mouse IgG. The experimental conditions are similar whether using affinity gels that are protein specific, i.e. antibody–antigen, or group specific, like IgG-Protein G chromatography. The protein sample is loaded onto the gel at an appropriate pH and usually at physiological salt concentrations. With certain proteins, however, high salt concentrations in the buffer promote binding. This is because affinity binding is usually due to some combination of ionic and hydrophobic interactions occurring simultaneously. The binding of protein to the gel is strong, and dilute protein samples can be applied to the column because of the concentrating effect of the gel. Following binding, there are two general ways to elute the protein sample from the column.

When the ligand used is specific for the particular protein, as in antibody–antigen interactions, a highly concentrated, free solution of the same ligand is used to elute the protein from the column. The ligand in free solution is applied at a much higher concentration than the gel-bound ligand, which by shifting equilibrium liberates the protein from the immobilized ligand. The protein and free ligand complex eluted from the column can be separated from each other (e.g., dialysis) in a downstream step. The method of elution with this type of affinity gel usually employs a single step gradient, as in the batch mode described previously.

Affinity column chromatography that utilizes group specific ligands requires alternative methods to elute the bound protein. For example, with Protein G affinity chromatography, it would be inconvenient, impractical, and expensive to elute the IgG with a concentrated solution of Protein G; therefore, elution is usually accomplished by a pH change. Either a step or a linear gradient can be applied, and the choice depends on the particular application. In some systems (e.g., Protein A), resolution of subtypes of bound proteins can be achieved using a linear pH gradient or a step gradient of small pH increments. Usually, the pH of the collected fraction is adjusted back to a value where the protein is stable immediately after elution. Resolution is not greatly affected by the same chromatographic parameters, as in ion exchange and hydrophobic interaction, and is almost solely dependent on the protein/ligand specificity. The various types of affinity binding are shown schematically in Figure 5.7.

A special category of affinity chromatography that has useful but limited applications is Immobilized Metal Affinity Chromatography, or IMAC. IMAC is useful for purifying proteins that either bind metal ions as part of their biological function (metalloproteins) or that contain a large number of histidine residues on their surfaces. The metal ion is

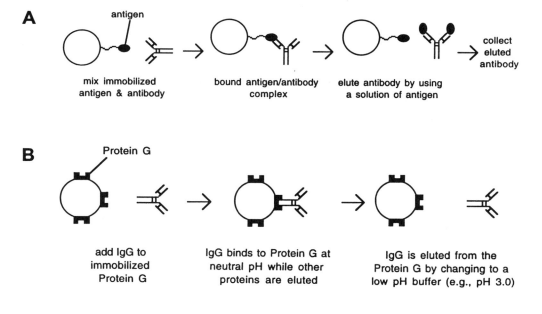

Figure 5.7 (A) Affinity isolation of an antibody using a gel bound antigen; (B) Affinity purification of IgG using gel bound Protein G.

immobilized to the gel, and the protein sample is loaded onto the column where it binds to the metal ion. The binding of the protein to the metal ion occurs via a metal specific binding cavity, as with metalloproteins, or through histidine residues on the protein surface. The bound protein is usually released from the gel by eluting the column with a concentrated solution of the same metal ion bound to the IMAC gel.

Chromatofocusing

Chromatofocusing is an extremely powerful technique in which the protein is purified based on its pI value. The technique is similar to isoelectric focusing gel electrophoresis, but it has the added advantage of high capacity and simple collection of the purified protein. The protein is applied to a weak ion exchange column and is eluted with a special type of buffer called Polybuffer®. The Polybuffer creates a linear pH gradient, and the protein elutes at the pH that matches its pI value. The protein elutes in a highly concentrated band due to the fact that the column has a focusing effect rather than the normal effect of dilution of the protein as it traverses the column.

5.3   EXPERIMENTAL DESIGN AND PROCEDURES

The protein extract prepared from the batch ion exchange experiment from Chapter 4 contains numerous contaminating proteins and requires further purification. Column chromatography provides the needed resolution to separate α-galactosidase from its contaminants. The chromatography will employ the diethylaminoethyl Sephadex A-50 anion exchange gel at pH 7.5 in a Tris buffer.

The IEX purified enzyme will then be assayed for activity, and the purification table will be completed. The analysis of the enzyme will be performed in Chapter 6.

Desalting of Batch Purified α-Galactosidase

The application of ion exchange column chromatography requires that the sample is in a low ionic strength buffer. Unfortunately, the impure α-galactosidase prepared in Chapter 4 is in Tris buffer, pH 7.5, containing 0.5 M sodium chloride. The first step before the column ion exchange experiment is to lower the ionic strength, i.e., the sodium chloride concentration, of the α-galactosidase sample. The technique used is often referred to as buffer exchange or, desalting. The desalting step can be carried out in several ways including dialysis, ultrafiltration, or gel filtration. The sample could also be diluted to reduce the ionic strength, especially since the sample volume is not critical. Diluted samples, however, may require a pH adjustment.

The dialysis technique employs the use of dialysis tubing, which is composed of a semipermeable membrane that allows unimpeded diffusion of low molecular weight components across the membrane. The dialysis tubing, which is purchased dry, is prepared by treatment

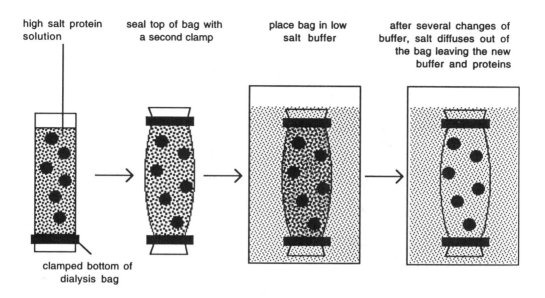

high salt protein solution

seal top of bag with a second clamp

place bag in low salt buffer

after several changes of buffer, salt diffuses out of the bag leaving the new buffer and proteins

clamped bottom of dialysis bag

Figure 5.8   Desalting by dialysis.

with a carbonate/EDTA buffer. It is then filled with the enzyme solution to be desalted. The ends of the tubing are tied off or clamped closed to form a dialysis bag, and the bag is then placed in a large volume of the new buffer. Dialysis tubing has a pore size that excludes all but the smallest proteins and salts from diffusing into the new buffer. In this way, the protein fraction of the sample is retained as the salt within the dialysis bag diffuses into the excess volume of the new buffer (Figure 5.8). Dialysis is generally conducted in a cold room with constant stirring and frequent buffer changes.

Actual fractionation of the sample by molecular weight can occur if dialysis tubing with a large pore size is employed. As the pore size of the tubing is increased, larger proteins are able to diffuse out of the dialysis bag, and, in this way, a partial purification by molecular weight can be accomplished.

Dialysis is a useful technique because large volumes can be desalted in one step; however, the drawbacks are that the technique is slow, and it requires several buffer changes. Due to the length of time needed for dialysis (usually 1–2 days with stirring), the entire process must be done in the cold to avoid any loss of biological activity.

Ultrafiltration is a desalting technique similar to dialysis. A semipermeable membrane is used to retain the protein component of the sample while buffer is pushed through the membrane (Figure 5.9). Typically, the protein solution is placed in an ultrafiltration apparatus which, when the pressure is applied, forces the solution through the membrane at the bottom of the vessel. The volume of the protein solution is decreased as liquid and salts pass through the membrane, but the larger molecular weight proteins are retained and concentrated. The protein solution is then diluted back to its original volume using the exchange buffer, and the process is repeated until the sample is effectively desalted. However, some protein loss may occur due to non-specific adsorption to the ultrafiltration membrane.

As in dialysis, choice of the membrane pore size can result in a limited purification based on protein molecular weight. Ultrafiltration is faster than the dialysis method but requires more elaborate and expensive instrumentation.

A third method for buffer exchange uses gel filtration chromatography. The separation of protein and salt is based on the large difference in

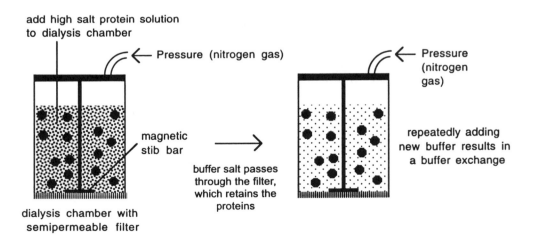

Figure 5.9 Mechanism of ultrafiltration.

their molecular weights. In the desalting application, the sample volume can be as high as 25% of the total column volume. For example, a 30 ml sample would require a total column volume of 120 ml of gel.

We recommend the use of gel filtration to desalt the α-galactosidase batch purified sample using disposable, prepacked columns such as the Pharmacia PD-10, Pharmacia HiTrap, or BioRad P6 desalting columns. These columns are generally inexpensive, reusable, and consistent. However, the drawback is that the sample volume that can be applied is small (≤ 2.5 ml); thus multiple sample applications are necessary.

Larger custom packed columns have the advantage that the entire sample can be desalted in one step; however, the column packing requires the purchase of a large column and a suitable pump to pack the column properly.

The use of the PD-10 column is described in the experimental section because, although the desalting process takes more time, using the PD-10 column is more convenient (i.e., they are prepacked and reusable) than preparing a desalting column. Whichever desalting column is used, simply follow the manufacturer's guidelines for its use.

Materials

PD-10 or other suitable prepacked desalting column
Clamps and ringstand
25 mM Tris-HCl, pH 7.5—Buffer A
10 ml disposable pipets
α-galactosidase batch IEX fraction Gal-1

Method

1. The PD-10 column comes prepacked with approximately 9.1 ml of Sephadex G-25, medium grade, stored in 20% ethanol. The cap on the column is removed, and the column is fitted onto a clamp and ringstand.

2. The tip of the column outlet is cut at the mark using a pair of scissors. The 20 % ethanol will begin to elute, but, due to the filter piece (frit) on top of the column, the buffer level will automatically stop when it drops to the level of the top filter. In this way, the column will not

accidentally run dry after the flow stops. This is an important feature of prepacked columns.

3. The column is equilibrated with 12 ml of ion exchange Buffer A (25 mM Tris-HCl, pH 7.5) prior to desalting. This flushes the storage buffer from the gel.

4. For convenience, focus on desalting only the Gal-1 fraction of highest activity obtained from the batch purification. Apply 2.5 ml of the Gal-1 sample to the top of the column. Allow it to enter the gel bed. At this time, the 2.5 ml of buffer eluting from the column can go to waste.

5. The protein is then eluted with 3.1 ml of Buffer A, and the column effluent is collected. The 3.1 ml sample collected in this step contains the desalted α-galactosidase. The desalting process is summarized in Figure 5.10.

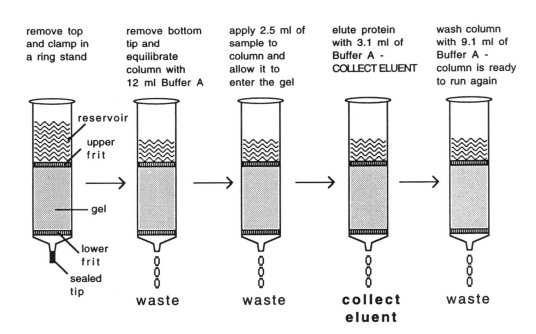

Figure 5.10   Desalting procedure with a PD-10 column.

6. The column is then washed with 9.1 ml of Buffer A. The sodium chloride from the initial 2.5 ml sample of Gal-1 elutes in this fraction and can be discarded.

7. Steps 1 through 6 are repeated until at least 10 to 15 units of enzyme activity are obtained.* The volume of Gal-1 necessary to obtain 10–15 units of enzyme activity is determined (as described in Chapter 4) by the following equations:

Total Units = (volume of Gal-1) × (Activity units/ml Gal-1)

so

Volume of Gal-1 = (Total Units)/(Activity units per ml of Gal-1)

8. Combine the desalted Gal-1 fractions and refrigerate until the column IEX experiment.

Preparation of the Ion Exchange Column

Although a prepacked column was used for the desalting of the Gal-1 sample, column chromatography often requires pouring a column. In this experiment, you will prepare a small ion exchange column which will be used for α-galactosidase purification.

Materials

Empty PD-10 (Pharmacia) or similar column—A PD-10 column is 1.5 cm in diameter and 8 cm long. The tip is 1 cm in length.

Clamps and ringstand

25 mM Tris-HCl, pH 7.5 (ion exchange Buffer A)

Plunger from a 10 ml syringe

DEAE Sephadex A-50 gel slurry

---

*Several columns in parallel can be used to speed up the processing of the sample.

Method

1. The column that will be used to pack the DEAE Sephadex A-50 ion exchange gel is an empty Pharmacia PD-10 column or equivalent. Each empty column comes with a cap and two plastic frits for the bottom and top of the gel. Graduate the column by filling with water in 1 ml increments. Mark the 1 ml divisions on the outside of the column up to 5 ml.

2. The bottom tip of the column is snipped off using scissors, and the column is mounted on a ring stand. Make sure the column is clamped securely and is vertically level.

3. The bottom frit is soaked in ion exchange Buffer A and is inserted into the bottom of the column. The buffer saturated frit is pushed to the bottom of the column using the plunger from a ten ml syringe. The rubber tip at the end of the plunger is first removed to avoid creation of a vacuum inside the column as the frit is pushed to the bottom (Figure 5.11).

4. The column is then filled with Buffer A, and the buffer is pushed through the column with the plunger from a ten ml syringe (with the rubber tip on) to initiate flow. In this case, the rubber tip is left on the plunger, and once the flow starts, the plunger is removed, and the buffer is allowed to drain out of the column. When all the buffer has been eluted, the bottom of the column is capped off.

5. Mix 4 ml of Buffer A and 4 ml of the DEAE Sephadex A-50 gel slurry and add this slurry to the column. Allow the gel to settle and then unplug the bottom of the column. Elute the column until the buffer meniscus (air-buffer interface) is a few mm above the line of the gel–buffer interface (Figure 5.11). Column elution is stopped by capping the column outlet. The gel–buffer interface can be difficult to see because of the transparency of the swollen gel. It may help to place a piece of colored paper behind the column to view the gel–buffer interface.

   Note: The DEAE Sephadex A-50 gel slurry from Chapter 4 may be used, or it is prepared the same way as described in Section 4.3.

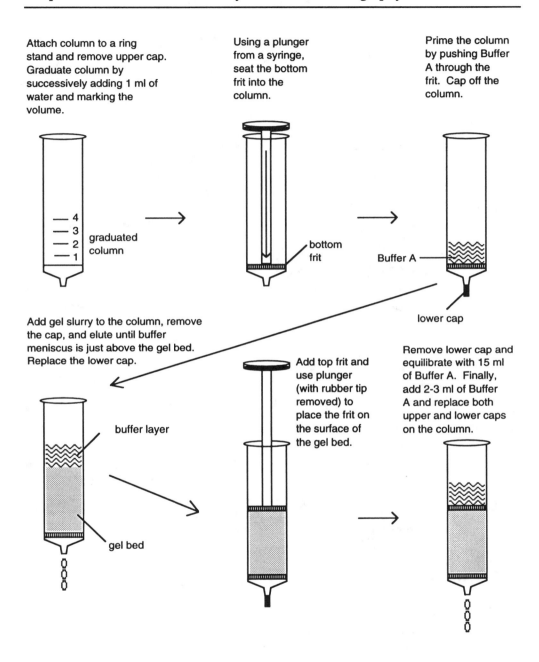

Attach column to a ring stand and remove upper cap. Graduate column by successively adding 1 ml of water and marking the volume.

Using a plunger from a syringe, seat the bottom frit into the column.

Prime the column by pushing Buffer A through the frit. Cap off the column.

graduated column

bottom frit

Buffer A

lower cap

Add gel slurry to the column, remove the cap, and elute until buffer meniscus is just above the gel bed. Replace the lower cap.

Add top frit and use plunger (with rubber tip removed) to place the frit on the surface of the gel bed.

Remove lower cap and equilibrate with 15 ml of Buffer A. Finally, add 2-3 ml of Buffer A and replace both upper and lower caps on the column.

buffer layer

gel bed

Figure 5.11 Preparation of a DEAE-Sephadex ion exchange column.

6. Add approximately 7 ml of Buffer A to the column, being careful not to disturb the gel bed.

7. Saturate the top plastic frit with Buffer A. Place the saturated frit onto the top of the column and evenly push it a few millimeters into the column.

8. Take the plunger from the 10 ml syringe (rubber tip removed) and in one stroke push the frit down the column, through the buffer, to the top of the gel bed. Remove the cap from the bottom of the column and elute the column with a total of 15 ml of Buffer A.

9. Cap off the bottom of the column just as the buffer from the previous step reaches the top frit. Add 2-3 additional ml of Buffer A and, using the plunger without the rubber tip, push the top frit approximately 0.5 ml into the bed to compress the gel. Be careful not to press the frit into the gel so much as to dry the bed. Use the volume guide that had been previously marked onto the column as a guide. The final volume of the packed column should be roughly 3–3.5 ml of gel.

10. Remove the cap at the bottom of the column and equilibrate the column with another 15 ml of buffer A. Note that the column flow will stop automatically when the buffer in the column reservoir above the gel reaches the top frit.

11. The column can be used immediately or stored in the refrigerator. For storage, the column outlet should be capped, a few ml of Buffer A is added to the column, and then the top of the column is capped. The column can now be stored in the refrigerator. For long term storage, the buffer can be made to 0.002% with chlorohexidine or stored in 0.05% trichlorobutanol.

Chromatography of the Desalted Protein Sample

Most ion exchange chromatography makes use of chromatography systems that automate much of the purification process. This instrumentation allows for reproducibility in the purification, especially with gradient formation. Unfortunately, chromatography systems generally cost tens of thousands of dollars. A cost effective alternative useful for instructional purposes employs ion exchange with step gradients. The ion exchange column used will contain the DEAE Sephadex A-50 gel and employ a step gradient. This will provide sufficient purification of the α-galactosidase while requiring a minimum of chromatography equipment. Before the specifics of this experiment are examined, an example of a purification using a chromatographic system is provided for comparison.

Ion exchange chromatography using an automated system requires a pump, a gradient maker, a packed ion exchange column containing a high performance IEX gel, an on-line detection system, and a fraction collector. A typical ion exchange run uses a linear gradient of either increasing salt concentration or pH using a typical system (Figure 5.12). The desalted Gal-1 sample can be purified using a linear gradient if the proper equipment is available. An example chromatogram of the purification of the desalted Gal-1 fraction using a linear gradient of increasing sodium chloride concentration is shown in Figure 5.13.

The sample is applied to the column using a one ml injection loop; however, the sample volume can be increased depending on the activity/ml of the sample. The column used in this experiment is a Q type ion exchanger (1 ml total volume) supplied prepacked by Bio-Rad. The designated buffers used for this purification gradient are: (A) 25 mM Tris-HCl, pH 7.5, and (B) 25 mM Tris-HCl, pH 7.5 + 0.5 M NaCl.

The column is first equilibrated with five column volumes of Buffer A (flow rate of 1 ml/min). The sample (in Buffer A by desalting on a PD-10) is applied to the column, washed, and then eluted with buffer by applying a linear gradient. The column begins elution with 100% Buffer A which is slowly replaced by addition of Buffer B. After 10 column volumes (10 ml) have been passed through the column, the buffer composition is now 100% Buffer B. This corresponds to a linear gradient from 0% Buffer B to 100% Buffer B over a total gradient volume of 10 ml (or 10 min). The column effluent is measured at 280 nm using an on-line detector and 0.5 ml fractions are collected.

The gradient trace is obtained using an on-line conductivity monitor which measures the increase in ionic strength of the eluting buffer. Immediately following the conductivity measurement fractions are collected and analyzed for α-galactosidase activity. The shaded area on the chromatogram corresponds to the fractions containing the α-galac-

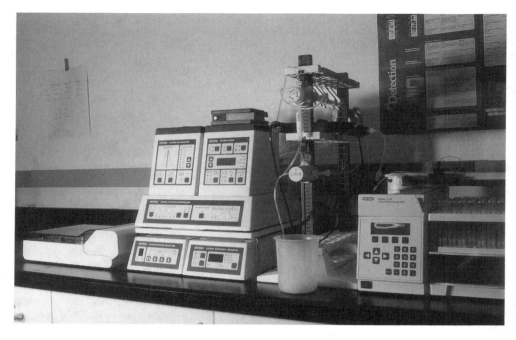

Figure 5.12 A representative chromatography system (the Bio-Rad Econo System) used for protein purification.

tosidase activity. Note that the α-galactosidase activity elutes at approximately 0.1 to 0.2 M sodium chloride concentration.

This chromatography experiment can be simplified by using a step gradient instead of a linear gradient. The step gradient is an increasing sodium chloride concentration gradient, but the gradient is formed by eluting the column with buffers of increasing ionic strength in discrete steps instead of continuously. The column used for this exercise is the ion exchange column that was previously packed with DEAE Sephadex A-50. The gel is beaded and has good flow properties, i.e., the large bead size results in low back pressure so that the buffer can pass through the column via gravity. The column is packed so that the flow automatically stops when the buffer reservoir is empty. In this way, the column cannot run dry which would create air pockets within the gel and ruin the packing.

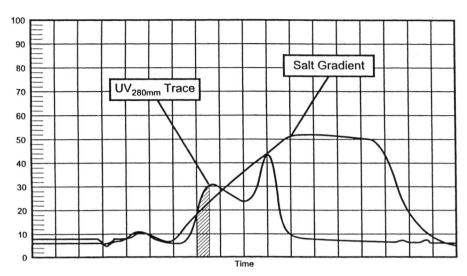

Figure 5.13   An IEX chromatogram of crude α-galactosidase using a linear salt gradient.

The sample, referred to as desalted Gal-1, is applied under low ionic strength conditions so that the proteins in the sample all bind to the gel. As the step gradient is applied, the various proteins are eluted. The fractions are manually collected and then analyzed for both α-galactosidase activity using the enzyme assay and for total protein content using the Lowry protein assay. The results are tabulated in the purification table.

Materials

DEAE Sephadex A-50 IEX column (prepared previously)

25 mM Tris-HCl, pH 7.5 (Buffer A)

25 mM Tris-HCl and 0.5 M NaCl, pH 7.5 (Buffer B)

13 × 100 mm test tubes

10 ml disposable pipets

Constant temperature block or water bath

0.5 M sodium acetate, pH 4.5

0.1 M sodium carbonate buffer

MORE...

Table 5.1	Elution Buffers for IEX Column		
Elution Buffer	Buffer A* (ml)	Buffer B** (ml)	[NaCl] (M)
1	5.4	0.6	0.05
2	4.8	1.2	0.10
3	4.2	1.8	0.15
4	3.6	2.4	0.20
5	3.0	3.0	0.25
6	2.4	3.6	0.30
7	1.8	4.2	0.35
8	1.2	4.8	0.40
9	0.6	5.4	0.45
10	0.0	6.0	0.50

*Buffer A = 25 mM Tris-HCl, pH 7.5
**Buffer B = 25 mM Tris-HCl pH 7.5 + 0.50 M NaCl

0.1 M p-nitrophenyl-α-D-galactoside, in water

Spectrophotometer

Micropipettes and tips

Method

1. Equilibrate the DEAE Sephadex A-50 ion exchange column with 15 ml of Buffer A (25 mM Tris-HCl, pH 7.5). If the column was just packed and equilibrated, this step can be skipped.

2. The desalted Gal-1 sample containing at least 10–15 units of α-galactosidase is applied to the DEAE Sephadex A-50 ion exchange column. The volume of the sample is not important but will probably be about 5–10 ml. The sample can be applied in more than one step.

3. Prepare a set of 10 elution buffers using the following combinations of ion exchange Buffers A and B shown in Table 5.1. These will be used for a step gradient elution of the α-galactosidase sample.

4. When the buffer level of the equilibrated column is at the top of the upper frit, Elution Buffer 1 is added (6 ml), and the effluent is collected in three 2 ml fractions. This can be done by adding 2 ml of water in the collections tubes and marking the level on the outside of the tube. (**Note:** Remove water before using collection tubes.) The tubes are labelled 1a, 1b, and 1c for the fractions from Buffer 1. Remember to keep the fractions refrigerated when not being used.

5. Repeat step 4 for each of the remaining nine elution buffers. Label each tube in a similar fashion as above. The experiment should generate thirty 2 ml samples.

6. The tubes labelled 1a–10a (ten tubes total) are each analyzed for α-galactosidase activity. For each tube that generates activity, the corresponding tubes labelled b and c are also analyzed for enzyme activity. For example, if no activity was found in tube 7a, then there is no need to analyze 7b and 7c. In this way, time and materials will be conserved by avoiding unnecessary analysis. Record the activities in a data sheet, e.g., as is shown in Figure 5.14.

α-Gal Assay

Elution Buffer (M)	Fraction	a	b	c
0.05	1			
0.10	2			
0.15	3			
0.20	4			
0.25	5			
0.30	6			
0.35	7			
0.40	8			
0.45	9			
0.50	10			

Figure 5.14   Data sheet for IEX purification.

Analysis of IEX Purified α-Galactosidase with the Lowry Protein Assay

The total protein content of the α-galactosidase fractions purified by ion exchange column chromatography must be determined so that the specific activities can be assessed. At this stage, the total protein content of the sample is too low for the Biuret assay to detect so the Lowry method is used instead. As discussed in Chapter 2, the Lowry protein assay is similar to the Biuret assay but is more sensitive.

A standard curve for the Lowry assay must be constructed for this measurement. This is similar to the Biuret standard curve experiment in Chapter 2. However, the Lowry assay will use much less of the protein sample, and the procedure is more complex due to the sensitivity of the assay and the instability of the Lowry reagent.

Materials

Reagent A—2% sodium carbonate in 0.1 N sodium hydroxide solution

Reagent B1—0.5% copper sulfate pentahydrate solution

Reagent B2—1% sodium tartrate solution

Reagent C—combine Reagents A, B1 and B2 in a ratio of 100:1:1

1 N Folin & Ciocalteu's Phenol Reagent (Sigma F 9252)—diluted with water from 2 N stock solution

13 × 100 mm test tubes

Bovine Serum Albumin, 0.5 mg/ml solution

Bovine Serum Albumin, unknown concentration (optional)

Spectrophotometer

Method

1. Use previously diluted standards or prepare a 0.5 mg/ml stock solution of Bovine Serum Albumin by dissolving 10 mg of BSA in 20 ml of deionized water. This stock may be stored at –20°C. Prepare at least 40 ml of Reagent C and 3 ml of diluted Folin & Ciocalteu's Phenol Reagent.

2. Use the previously constructed standard curve or prepare a new curve. To construct a new curve, label six (6) 13 × 100 mm test tubes #1 through 6. Add 400 μl, 300 μl, 200 μl, 100 μl, 50 μl, and 25 μl of the BSA stock solution to tubes 1 through 6, respectively. Add 0 μl, 100 μl, 200 μl, 300 μl, 350 μl, and 375 μl of deionized water to each tube to bring the total volume up to 400 μl. Label a seventh tube as the blank and add 400 μl of deionized water to this tube. Label an eighth tube "unknown" and add to it 400 μl of a sample of unknown BSA concentration. Using 400 μl of the fractions containing enzyme activity collected during the ion exchange experiment, also measure the total protein using this Lowry protein assay. As a control, also analyze the Tris elution buffer and subtract the value from the ion exchange fractions.

3. Add 2 ml of reagent C to each tube and incubate for 10 minutes.

4. Add 200 μl of 1 N Folin & Ciocalteu's Phenol Reagent to each tube. The Folin & Ciocalteu's Phenol Reagent is unstable, and the mixture must be vigorously mixed as the reagent is being added. Let the tubes incubate for exactly 30 minutes.

5. After 30 minutes, zero the spectrophotometer at 550 nm using the blank and then measure the absorbance of each tube.

6. For the protein standards, plot the absorbance on the y axis against mg protein on the x axis.

7. Measure the absorbance of the unknown BSA sample (optional) and the ion exchange fractions. Remember to subtract the absorbance reading of the Tris elution buffer from the absorbance of the IEX fraction. Determine where the absorbance falls on the standard curve and the corresponding protein mass of the samples.

8. The specific activity of each tested sample is calculated by dividing the enzyme activity by the corresponding total mg protein content of the sample.

9. The fraction that gives the highest specific activity is used to complete your purification table for the process and will be used for analysis in Chapter 6.

STUDY QUESTIONS

1. Propose an alternative procedure for column chromatographic purification of α-galactosidase if the enzyme has been batch isolated using ammonium sulfate precipitation. Explain your rationale.

2. Based on the chemical composition of α-galactosidase, propose an affinity technique that could be used to purify the enzyme instead of ion exchange column chromatography.

3. Propose a non-affinity column chromatographic technique for the purification of α-galactosidase other than ion exchange column chromatography. (Hint: The peptone in the culture media is a mixture of low molecular weight proteins.)

FURTHER READINGS

Deutscher M, ed. (1990): *Methods in Enzymology: Guide to Protein Purification*, Vol. 182, San Diego: Academic Press

Janson J, Ryden L, eds. (1989): *Protein Purification, Principles, High Resolution Methods and Applications*. New York: VCH Publishers

6

Protein Analysis and Verification

6.1 OVERVIEW

The analysis of a protein is usually an attempt to characterize its structure and/or determine its degree of purity. Proteins of unknown structure are purified to apparent homogeneity and then analyzed to determine structure. The success of the structural analysis is, of course, critically dependent on the level of purification achieved.

If the structure of the protein is already known, then the analysis is performed to determine the level of purity, and it is, therefore, simply a critique of the method(s) used for the purification. To give an example of this situation, when a company wishes to test an improved method it has developed for the production of a currently manufactured protein whose structure is well known, the purity of the protein is analyzed simply to evaluate the new purification process. The improved purification process could lead to increased profits either due to increased product purity or to improved process throughput.

The various methods employed for the analysis of proteins will be discussed in the background section along with an assessment of the relative importance of the various techniques. Overlap of some material previously discussed is unavoidable because methods used for preparative purification of proteins (e.g., chromatography) can also be used to analyze the protein for impurities.

The purification scheme for α-galactosidase has progressed through several steps, including its induced synthesis, batch capture, and column purification. This isolation and purification of α-galactosidase has been an attempt to simulate an actual purification because we have not assumed any prior knowledge of the enzyme structure. The exercises described in this chapter represent the last step in our purification process. The IEX purified α-galactosidase will be analyzed by native polyacrylamide gel electrophoresis. The purpose of the analysis is to evaluate the purity of the enzyme and also to be the final purification by extracting the enzyme from the gel (if the extraction step is determined to be necessary).

6.2   BACKGROUND

The assessment of purity for a protein can be ambiguous due to the relative nature of the word purity. A protein preparation may be heterogeneous with respect to its mobility in an isoelectric focussing gel due to glycosylation, but it may be homogeneous with respect to molecular weight analysis and biological activity. The protein purity greatly depends on the criteria which are chosen to define homogeneity.

Methods used for protein analysis typically incorporate both biological and chemical criteria. Some of the more widespread methods available for analysis of proteins are summarized in Table 6.1.

Gel Electrophoresis

Gel electrophoresis is one of the most powerful methods for protein analysis that is currently available. Recent advances in automation, commercial availability of precast electrophoresis gels, and staining protocols have made electrophoresis a routine laboratory technique.

In practice a protein sample is applied and separated onto a gel matrix, typically polyacrylamide cross-linked with bis-acrylamide, which is cast

Table 6.1 Methods Used for Protein Analysis

Technique	Protein Property Exploited
Chromatography	
Ion Exchange	charge
Gel Filtration	size, subunits
Hydrophobic Interaction	hydrophobicity
Reversed-Phase	hydrophobicity
Chromatofocusing	pI (isoelectric point)
Affinity	biological activity
Electrophoresis	
Native Gel	mass/charge
Denaturing Gel (SDS-PAGE)	molecular weight/subunits
IEF	pI (isoelectric point)
Spectrometry	
Mass Spectrometry	molecular weight
Composition Analysis	
Amino Acid Analysis	amino acid ratios (content)
Sequence Analysis	primary structure

as a square slab gel with a thickness between 0.5 to 2.0 mm. The protein sample is introduced into the polyacrylamide slab gel, which acts as a molecular sieve when an electric field is applied. The protein migrates through the gel, and depending on the conditions of the gel and buffers, separation of the protein components will occur. The proteins are visualized within the gel following electrophoresis usually by some type of staining that interacts with the protein. Normally, the entire gel is stained and then destained, leaving a transparent gel and colored protein bands.

Traditionally, gel electrophoresis has been considered an analytical technique because the method and/or the detection is destructive to the protein sample (i.e., loss of activity). Recent advances in blotting technology, i.e., the transfer of the protein from the gel to a membrane, have made some electrophoresis techniques preparative in the sense that the separated proteins retain their biological activity and can be removed from the gel for further analysis or experiments. The scale, however, is small when compared to chromatography.

Native PAGE (PolyAcrylamide Gel Electrophoresis) separates proteins under native conditions that are nondenaturing. The proteins are separated by a combination of their charge and mass. The greater the charge on the protein, the faster the protein migrates in the electric field. The larger the protein, the more its velocity is slowed by the sieving effect of the gel. In addition to being an excellent analytical technique, proteins can be purified in small quantities by this technique due to the nondenaturing conditions which enable the protein to be extracted from the gel while retaining biological activity.

SDS-PAGE (Sodium Dodecylsulfate–PolyAcrylamide Gel Electrophoresis) is an example of a denaturing electrophoresis technique. The method is normally used as a technique of analysis, but it also has limited preparative applications. The protein is denatured and reduced by treating it with a sample buffer containing detergent, typically SDS, and a reducing agent such as β-mercaptoethanol. The sample is then heated to 90–100°C for five minutes to insure complete denaturation. The SDS binds to the denatured proteins and evenly coats all the proteins with a negative charge. The SDS possesses a negative charge; thus SDS evens the charge distribution between large and small proteins so that all the proteins have the same charge to mass ratio. Since the proteins all contain the same charge to mass ratio, the migration of the proteins down the gel is solely a function of their molecular weights.

Aside from its use in determining protein homogeneity, SDS-PAGE is also used to determine the molecular weights of unknown proteins. In this manner, SDS-PAGE is similar to gel filtration in which the proteins are separated by their size and shape (with globular proteins, size is a function of molecular weight). A standard curve is generated from a mixture of known proteins of various molecular weights, which is used to determine the molecular weight of the unknown protein. Additionally, SDS-PAGE provides information on the subunit structure of a protein due to the denaturing and reducing environment in the sample buffer.

IsoElectric Focusing (IEF) is an extremely powerful electrophoretic technique that separates proteins on the basis of their pI (isoelectric point). The gel is prepared using special buffers called ampholytes, and when an electric field is applied to the gel, the ampholytes migrate and form a pH gradient. A protein applied to the gel will migrate until it reaches the pH that matches its pI. At that point, the protein loses its charge and stops migrating. If the now neutral protein diffuses in the gel, it immediately becomes charged again and migrates back to its pI. In this way, the protein is focused, which results in a separation with extremely high resolution. IEF is a nondenaturing technique, and the protein can be recovered from the gel. However, the protein must be separated from the ampholytes after it is extracted.

Chromatography

Up to now chromatography has been discussed as a preparative technique, but it can also be used analytically to determine the purity of a protein. Unlike preparative purification, special instrumentation, such as HPLC, is usually employed for analytical chromatography. HPLC stands for high performance liquid chromatography and is used both analytically and preparatively (i.e., when relatively small amounts of protein need to be purified).

Analytical chromatography involves monitoring the elution profile of the protein sample, usually by UV absorbance. The sample, which is typically in μg amounts, is not usually recovered. Preparative chromatography usually involves physically recovering the protein sample as it elutes from the column. Preparative biocompatible high performance chromatography, such as FPLC® (Pharmacia), is routinely applied to the purification of proteins. With biocompatible LC systems, instrumentation and prepacked columns are either glass, plastic, or inert titanium, i.e., materials that are biocompatible and do not denature the protein. Protein micropreparative chromatography is used when purifying and recovering very small amounts of bioactive protein; it differs from traditional analytical chromatography.

There are many high performance columns commercially available that are highly efficient and result in excellent resolution. Analyses typically rely on a chromatographic technique with a different chemical selectivity than that used for the preparative purification step. For example, a protein purified on a preparative ion exchange column (separation by charge) would be analyzed by a complementary technique such as reversed phase chromatography (separation by hydrophobic interaction).

Due to its high resolution, reversed phase chromatography is a popular technique used to determine the purity of a protein, or peptide. Peptides can be purified and isolated because they typically do not have tertiary structure and are not denatured under the reverse phase conditions. Proteins can be analyzed by reverse phase but are usually not recovered with biological activity intact. Reversed phase conditions normally make use of organic solvents that denature most proteins.

High performance ion exchange, hydrophobic interaction, and gel filtration columns are commercially available. Since these columns are nondenaturing techniques, they can be used for both analytical and preparative chromatography.

In addition to gel filtration and electrophoresis, molecular weight analysis of proteins can be accomplished by sedimentation rates of the protein using an ultracentrifuge and, more recently, by mass analysis using mass spectrometry. Though currently limited in its application, mass spectrometry is valuable because it directly measures the molecular weight of the protein and not the size, shape, or charge.

Amino Acid Analysis

The determination of amino acid composition and sequence are usually the first chemical analyses performed on a purified protein. In amino acid analysis, the protein is hydrolyzed, usually in six molar hydrochloric acid overnight at 110°C, and the amino acid content is determined using an amino acid analyzer. The hydrolysis degrades the protein into its constituent amino acids which the analyzer separates and quantitates. Commercially available amino acid analysis instrumentation separates the amino acids by either ion exchange or reversed phase chromatography. Amino acid analysis gives the ratios of the constituent amino acids but gives no information about their sequence. The data are usually presented as a table containing the various ratios of the amino acids (Table 6.2).

Protein Sequencing

The most common method for sequencing a protein is by automated Edman analysis, which is the chemical technique used in most commercial instrumentation currently available on the market (Figure 6.1).

Table 6.2 A Representative Amino Acid Analysis Data Sheet From the Hydrolysis and Analysis of α-Galactosidase.

Amino Acid	grams amino acid 100 grams protein	residues calculated for 52,000 dalton protein
ala	4.7	34
arg	5.1	17
asp (asn)	15.7	71
cys	2.0	10
glu (gln)	7.7	31
gly	5.0	46
his	1.9	7
ile	5.0	23
leu	8.0	37
lys	4.7	19
met	3.3	13
phe	5.9	21
pro	3.2	17
ser	7.3	43
thr	4.9	25
trp	4.3	12
tyr	7.8	25
val	3.8	20

The derivatized amino acids are sequentially removed from the protein chain and then analyzed by reversed phase chromatography, and in this way, the sequence of the protein can be determined. The Edman technique is limited in that the analysis only works reliably for short chain polypeptides (e.g., less than 30–40 amino acids). Sequencing of the purified protein can be done either by analysis of the total protein or by limited analysis of a partial sequence. The sequence of the entire pro-

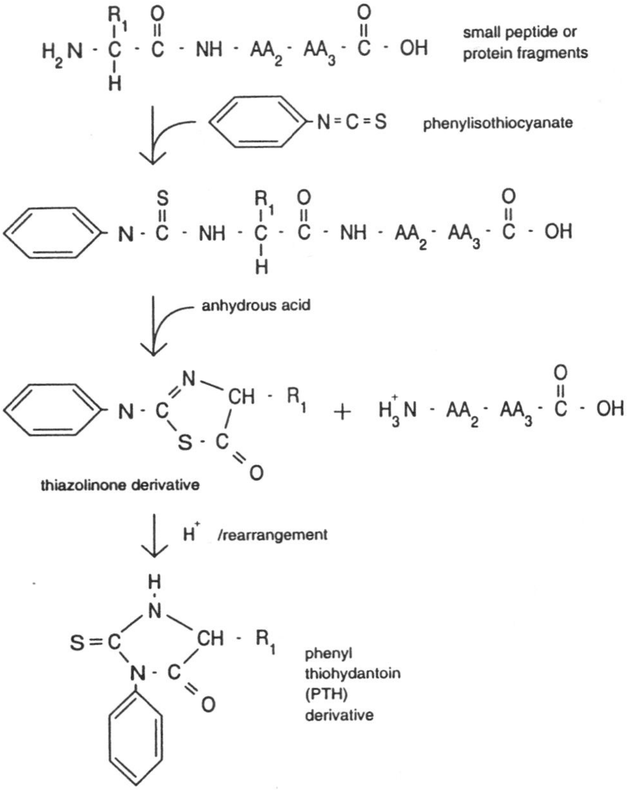

Figure 6.1 Edman chemistry for amino acid sequencing.

tein chain can be determined by first generating smaller polypeptides using a combination of enzymatic and/or chemical cleavage and then analyzing each fragment. The individual polypeptide fragments are separated either chromatographically or electrophoretically (see below) and analyzed on a protein sequencer. In this way, the sequences of the individual fragments are elucidated, and then, working backwards, the complete amino acid sequence is determined.

A more practical and convenient procedure than total sequence analysis is limited sequence analysis of the N-terminus of the protein chain. The information provided by a limited sequence analysis is generally sufficient to design a small oligonucleotide probe that can be used to find the gene that codes for the protein. The sequencing of DNA is much easier to perform than total protein sequencing. Therefore, it may be more efficient to find the gene that codes for the protein and then clone and sequence the gene in order to determine the sequence of the protein. However, the nucleotide sequence of a gene does not necessarily reflect modifications that can occur to the protein posttranslationally.

6.3 EXPERIMENTAL DESIGN AND PROCEDURES

The enzyme has been purified by ion exchange chromatography and is now ready for analysis. Since the α-galactosidase will ultimately be used for protein sequencing, its purity requirements are rather stringent. To ensure purity, evidence needs to be based on both biological (i.e., high specific activity) and chemical (purity based on molecular weight, charge, etc.) criteria.

The following analysis will use native polyacrylamide gel electrophoresis coupled to several methods for enzyme detection. Native PAGE is well suited for this analysis since it can lead to both biological and chemical measurements.

The purpose of this analysis is to demonstrate purity and is potentially the final step in the purification process. The enzyme preparation derived from the ion exchange column is further fractionated by native PAGE, and the detection of a single enzyme band indicates that the enzyme is pure (based on the electrophoretic criteria). Although the actual experiment is not performed here, the purified α-galactosidase would subsequently be subjected to sequence analysis, and the results would be used to design an oligonucleotide probe for identifying the gene that encodes the enzyme.

Even if the electrophoresis of the sample indicates heterogeneity by revealing several bands, the band corresponding to α-galactosidase can be excised, isolated, and sequenced. In this way, the electrophoresis not only determines the purity of the sample but may also be the final purification step in the protocol.

The separation of the constituent proteins in a sample analyzed under native PAGE is based on their charge to mass ratio. In SDS PAGE, proteins are separated solely on the basis of their differences in mass, and, therefore, native gel electrophoresis may be somewhat more discriminating. More importantly, because the protein or enzyme being electrophoresed is in its native state, it can be visualized by methods that exploit the proteins' biological activity. Under denaturing conditions, only chemical methods of visualization are possible because the proteins analyzed are no longer biologically active. Visualization using biological activity is not only more sensitive than most chemical methods, but it also identifies which band is the actual protein of interest. For example, a chemical method of visualization will detect the α-galactosidase in addition to all other contaminating proteins (if any). If the activity of α-galactosidase is not detected, there is no way of knowing which band visualized in the gel is actually α-galactosidase. If we visualize the enzyme using the α-galactosidase assay only active enzyme will be detected. Unfortunately, this is not an absolute rule because sometimes more than one form of an enzyme (isozyme) can be biologically active, but nevertheless, detection of enzyme activity is still advantageous.

The use of native conditions is essential if the purified protein is used in a subsequent step that requires the protein to be biologically active. Therefore, the α-galactosidase preparation will be analyzed using native polyacrylamide gel electrophoresis and with both a chemical method (Coomassie Blue staining) and two types of enzyme assays for biological activity.

The electrophoresis will be performed using a 6% polyacrylamide resolving gel with a continuous buffer system (i.e., a single buffer system) and no stacking gel. A stacking gel is a dilute acrylamide solution which is often used to form the upper portion of the gel, including the wells. Proteins travel rapidly through the stacking gel which causes them to concentrate into tight bands on the stacking gel/resolving gel interface. The continuous gel system used is less complicated than discontinuous (multiple buffers) systems and is just as efficient in resolving large enzymes, such as α-galactosidase. We recommend the use of a Bio-Rad Mini PROTEAN® II electrophoresis system, but any mini gel system is appropriate.

Preparation of the Acrylamide Gel

A solution of acrylamide and bis-acrylamide (N,N'-methylene-bis-acrylamide) can be solidified (i.e., polymerized) into a gel matrix by the addition of ammonium persulfate and TEMED (Figure 6.2). This gel is a standard matrix used in the separation of proteins by electrophoresis.

A

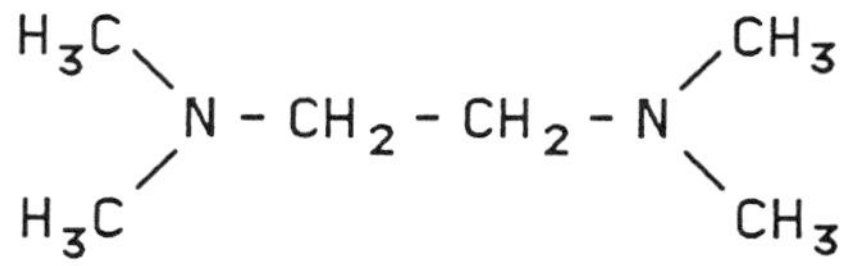

$$H_3C - N - CH_2 - CH_2 - N - CH_3 \text{ (with } H_3C \text{ and } CH_3 \text{ branches)}$$

N,N,N'N' tetramethyl ethylenediamine
(TEMED)

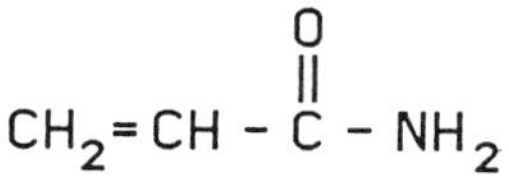

$$CH_2 = CH - \overset{O}{\overset{\|}{C}} - NH_2$$

acrylamide

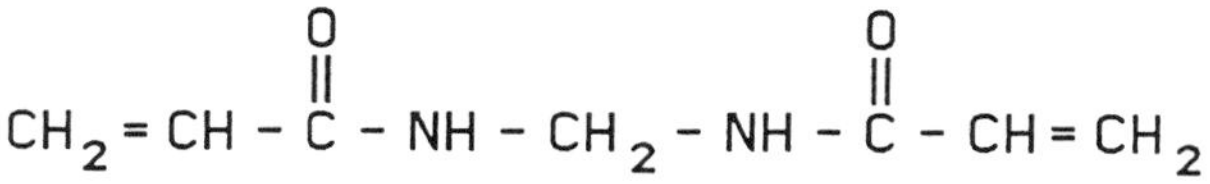

$$CH_2 = CH - \overset{O}{\overset{\|}{C}} - NH - CH_2 - NH - \overset{O}{\overset{\|}{C}} - CH = CH_2$$

N,N'-methylene-bis acrylamide

B

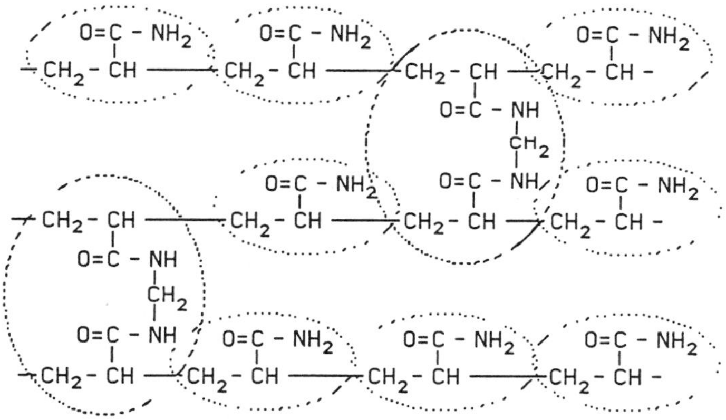

Figure 6.2 (A) The components of polyacrylamide. (B) Polymerization and cross-linking of acrylamide and N,N'-methylene-bis-acrylamide.

Materials

Tris-HCl

Boric Acid

Ethylenediaminetetraacetic acid, disodium salt (EDTA)

MORE...

40% acrylamide solution (Sigma A 6050)—A 37:1 ratio of acrylamide to bis-acrylamide.

TEMED

Ammonium persulfate

Method

1. Only prepare the acrylamide solution after the electrophoresis plates have been assembled (following experiment). The 6% acrylamide gel is prepared in a small sidearm flask using the following recipe:

> 6.5 ml of distilled water
> 2.0 ml of 5X TBE
> 1.5 ml of 40% acrylamide* (40% T, 2.6% C)
> 50 µl of 10% ammonium persulfate
> 10 µl of TEMED

Caution! Acrylamide is a neurotoxin and should be handled with extreme caution.

The 5X TBE is a buffer concentrate of Tris, borate, ethylenediaminetetraacetic acid, pH 8.4. The buffer is prepared by mixing 54.5 g Tris base (Mol Wt 121.1 g/mole), 24.75 g Boric Acid (Mol Wt 61.8 g/mole), and 4.65 g Ethylenediaminetetraacetic acid, disodium salt (Mol Wt 372.74 g/mole) in one liter of distilled water. The pH of the buffer should be 8.4.

The concentration of the acrylamide solution is 40% and is a 37:1 ratio of acrylamide to the bis-acrylamide cross-linker. The bis-acrylamide cross-linker is responsible for the three dimensional structure of the gel. Ammonium persulfate is a free radical source and initiates the polymerization reaction. The ammonium persulfate solution should be freshly prepared by dissolving 100 mg of ammonium persulfate in 1 ml of water.

---

*Acrylamide/bis-acrylamide solutions are often reported as "%T" and "%C." The "T" refers to Total combined percent of acrylamide and bis-acrylamide, while the "C" refers to the percent bis-acrylamide of the total. The 40% solution of 37:1 acrylamide/bis-acrylamide is reported as 40% T, 2.6% C. The 2.6% C is calculated by dividing 1 by 38 (i.e., the total of the ratio). Therefore, the 40% solution of acrylamide contains, per 100 ml, 37.4 g of acrylamide and 2.6 g of bis-acrylamide.

TEMED is the source of the stable free radicals that perpetuate the polymerization reaction. Oxygen dissolved in the acrylamide solution can scavenge the free radicals and inhibit polymerization; thus removal of the oxygen by degassing the acrylamide solution prior to the addition of TEMED and ammonium persulfate is usually necessary. The acrylamide, buffer concentrate, and water are mixed together and degassed by covering the top with a rubber stopper and applying a vacuum (some researchers consider degassing optional). The TEMED is then added to the solution with mixing. The polymerization reaction is initiated by addition of the ammonium persulfate solution with rapid stirring. The acrylamide solution is used immediately and will completely polymerize within one hour or less.

Casting the Gel

Due to the ease in which gels can be cast, we recommend using the Bio-Rad Mini-PROTEAN® II electrophoresis system; however any standard mini gel system is acceptable. If a commercially available system is used, the directions for casting the gel and assembly of the electrophoresis chamber should be in the instrument instruction manual. The directions presented here are appropriate for casting a minigel using any simple, generic vertical electrophoresis system (Figure 6.3).

Materials

Glass plates
Spacers for a fixed gel width (0.75–1.0 mm thickness)
Clamps—use clamp-like binder clips
Sealing wax—low melting paraffin sealing wax
Comb—with at least 9 wells per student/group

Method

1. Wash the plates with a glass cleaner and dry plates thoroughly before casting the gel. Make sure the plates are lint free.

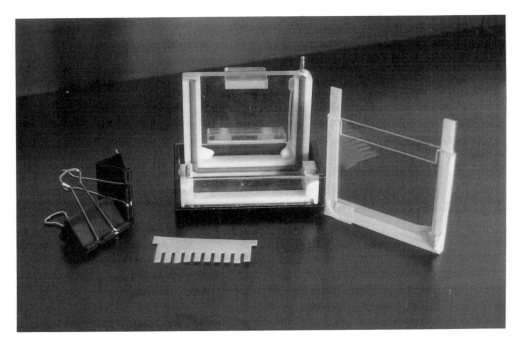

Figure 6.3   Components of a vertical electrophoresis system.

2. Place the front glass plate (plate with the greater vertical length) on the lab bench. The gel spacers are placed on top of the plate, two on each side and one along the bottom. All three spacers should be approximately 3–5 mm from the plate's edge. Make sure they are evenly aligned (Figure 6.4).

3. Place the shorter rear plate evenly on top of the longer glass plate without disturbing the spacers. The plates should be flush along the bottom. Clamp the glass plate sandwich with two clamps on the bottom and two clamps on each side.

4. Seal the three sides of the gel sandwich by applying melted wax with a Pasteur pipet. The hot wax is used to fill the channel formed along the edge of the glass plates and the spacer. The wax seals the glass plate sandwich and prevents it from leaking as the acrylamide solution is poured between the plates.

5. Place the plastic comb (that forms the sample wells) partially into the glass plate sandwich at an angle of about ten degrees. The glass plate sandwich is positioned vertically, and the acrylamide solution is poured between the plates using a Pasteur pipet. Apply the gel slowly so the acrylamide pours evenly without trapping any air bubbles. Fill the glass plate sandwich to the top of the shorter glass plate and fully insert the gel comb. Some excess acrylamide may run down the plate, which can either be absorbed with a paper towel or scraped off later. Allow the gel to polymerize undisturbed for approximately 45–60 minutes.

Assembling the Electrophoretic Buffer Chamber

The glass plates/acrylamide sandwich is secured into an electrophoresis chamber that has upper and lower buffer chambers. When electrical leads are connected to these chambers and a current is applied, the only route for the current is through the acrylamide gel. Proper assembly of this electrophoresis chamber is necessary to ensure that the current only travels its correct route.

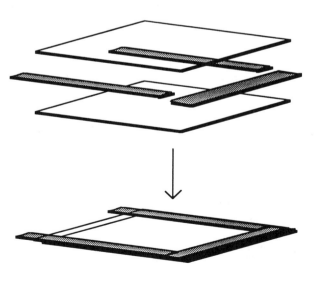

Figure 6.4 Glass plate sandwich for gel casting.

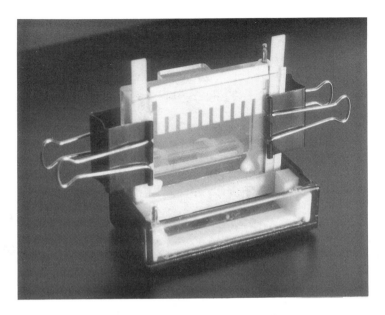

Figure 6.5   Clamping the gel sandwich into the buffer chamber.

Materials

Polymerized acrylamide gel
Vertical electrophoresis chamber
1X TBE

Method

1. Once the gel has polymerized, remove all the clamps and the spacer along the bottom of the gel.

2. The glass plate sandwich (containing the polymerized gel) is now clamped to the electrophoresis chamber with the shorter rear plate opening into the upper buffer chamber. The lower and upper buffer chambers are both filled with 1X TBE buffer. The buffer level in the lower reservoir must be above the gel line, and the upper reservoir must cover the sample wells of the gel (Figure 6.5).

3. Gently remove the comb from the gel. Take care not to tear the acrylamide separating the wells. Using a Pasteur pipette, flush out any bubbles trapped in the wells or along the bottom edge of the gel.

Sample Loading and Electrophoresis

The gel is essentially divided into three equal parts for the analysis of the purified α-galactosidase. Protein standards will be flanked on either side by a sample of the α-galactosidase purified by ion exchange chromatography in Chapter 5. (Use the fraction that has the highest specific activity; see Figure 6.6.) The protein standards are a series of five proteins covering a molecular weight range from approximately 650,000 daltons down to 65,000 daltons and can be purchased commercially. The purpose of the protein standards is for qualitative comparison to proteins resolved in the α-galactosidase sample. These standards cannot be used quantitatively because the separation of the proteins is by charge/mass ratio and not by mass alone as in SDS-PAGE.

The order of sample application to the gel is shown in Figure 6.6.

The gel will be electrophoresed at a constant voltage of 200 V for one hour. The gel is then removed from the apparatus but is left on the glass

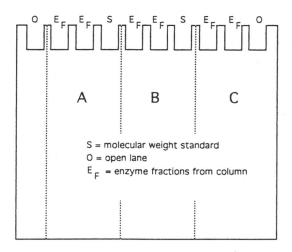

Figure 6.6 Sample application to gel.

plate to facilitate handling. The entire gel is first soaked in a 0.25 M sodium acetate buffer to prepare it for visualization by the enzyme assay techniques.

Materials

1X TBE

50 μl Hamilton Syringe or micropipette with gel loading tips

Protein standards—either purchased individually or as a kit (e.g., Pharmacia Electrophoresis Calibration Kit):

thyroglobulin	669,000 daltons
ferritin	440,000 daltons
catalase	232,000 daltons
lactate dehydrogenase	140,000 daltons
bovine serum albumin	67,000 daltons

Loading buffer—0.25% Bromophenol Blue, 40% sucrose

Constant voltage power supply

Method

1. The protein standards are prepared to a concentration of 1–2 mg/ml of each protein in water. Mix 80 μl of protein standard with 20 μl of loading buffer (use 1 μl of loading dye to 4 μl of sample). Load 25–30 μl of protein standard into the wells as depicted in Figure 6.6. Mix 160 μl of the α-galactosidase sample with 40 μl of the loading buffer and load 30–35 μl into the appropriate wells (Figure 6.6). The samples are applied to the gel using a 50 μl Hamilton syringe or micropipette with gel loading tips according to the schematic in Figure 6.6.

2. The lid of the electrophoresis unit is attached, and the electrodes are connected from the power supply to the electrophoresis chamber. The gel is run under a constant voltage of 200 V for approximately one hour. (**Note:** The tracking dye may run off the gel by this time.) Turn off the power and disconnect the leads.

3. The glass plate gel sandwich is removed from the assembly after the electrophoresis is complete. The spacers are removed, and then the

shorter rectangular glass plate is also removed by inserting a micro spatula (carefully!) between the glass plates and gently breaking the seal between the gel and the shorter rectangular glass plate.

4. The gel is left on the longer rectangular glass plate and the gel/plate is placed in 0.25 M sodium acetate buffer, pH 4.5 and equilibrated for approximately 10 minutes. The gel is equilibrated in this buffer in preparation for the subsequent visualization experiments.

Coomassie Blue R-250 Staining

The gel will then be cut into three equal pieces as shown in Figure 6.6. The first section, labeled "A," will be visualized by Coomassie detection. The Coomassie Blue dye stains both the gel and the proteins a deep blue. When the gel is subsequently destained, the protein bands retain the dye and remain as blue bands against the transparent gel background. The Coomassie detection is moderately sensitive and will detect a protein mass of approximately 1 µg. The Coomassie dye, however, is a chemical method of detection, and it does not provide any information concerning the identity of the visualized protein(s).

Materials

Coomassie Blue R-250
Methanol
Acetic acid

Method

1. The gel/plate is removed from the equilibration buffer and cut into three sections with a razor blade (Figure 6.6).

2. The gel section "A" is removed from the glass plate and is Coomassie stained in a small plastic weigh boat (or other suitable container) for one half hour. The staining solution is 0.1% Coomassie Blue R-250, 40% methanol, 50% water, and 10% acetic acid.

Note: Filter the freshly prepared staining solution before use.

3. The gel is then destained in a solution of 40% methanol, 50% water, and 10% acetic acid until the background gel is transparent. The destaining will require 1–2 hours with frequent changes of destaining solution. Gels can be destained overnight if necessary.

4. The distances that the protein bands migrate from the bottom of the wells are measured (in mm) and divided by the distance traveled by the tracking dye (or total gel length). This value is used to report the relative migration of the particular protein bands.

relative migration = distance in mm / total length of gel in mm

Visualization by Assay with p-Nitrophenyl-α-D-galactoside

Section "B" of the gel will be analyzed using the α-galactosidase assay which employs the substrate p-nitrophenyl-α-D-galactoside. Using a razor blade, the section of gel is cut into three lanes. The lane corresponding to the enzyme sample is cut into 0.25–0.5 cm strips (Figure 6.7).

The gel slices are then placed into a series of test tubes and assayed for enzyme activity. A positive assay determines the location of the enzyme in the gel. The distance migrated by the enzyme can be measured and compared to that migration distance of the protein standards that were visualized by the Coomassie blue.

Materials

0.5 M sodium acetate, pH 4.5

0.1 M p-nitrophenyl-α-D-galactoside, in water

0.1 M sodium carbonate buffer

13 × 100 mm test tubes

Microspatula

Razor blade

Micropipettes and tips

Spectrophotometer

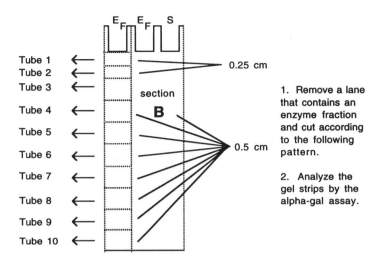

1. Remove a lane that contains an enzyme fraction and cut according to the following pattern.

2. Analyze the gel strips by the alpha-gal assay.

Figure 6.7 Slicing of gel for enzyme assay.

Method

1. Section "B" of the previously sliced gel is used in this assay. Lane E_F of section "B" is cut into slices as indicated in Figure 6.7. The unused portion of the gel will serve as a backup if necessary. For convenience, the gel is left on the glass plate at this point. Each gel slice is placed (carefully) into the bottom of a 13×100 mm test tube labeled one through ten, respectively. Add 100 μl of 0.5 M sodium acetate buffer, pH 4.5, and 50 μl of water to each tube.

   Note: Depending on the migration distance of the enzyme, it may be advantageous to make the first two slices 0.25 cm and the subsequence slices 0.5 cm thick.

2. The gel slices are fully submerged into the liquid, and then 50 μl of p-nitrophenyl-α-D-galactoside is added to each tube.

3. The tubes are allowed to incubate for one hour at room temperature.

Tube	Distance Migrated into Gel (cm)	α-Gal Activity
1	0.25	
2	0.50	
3	1.00	
4	1.50	
5	2.00	
6	2.50	
7	3.00	
8	3.50	
9	4.00	
10	4.50	

Figure 6.8   Data presentation of gel slices.

4. Each tube is quenched by adding 3 ml of a 0.1 M sodium carbonate solution. A positive result for the presence of α-galactosidase is the appearance of a yellow color due to the hydrolysis of the substrate. The tubes can be analyzed quantitatively in the spectrophotometer if desired. The data can be summarized as shown in Figure 6.8.

5. The distance α-galactosidase migrated is determined by the gel slice. Divide the distance migrated (mm) by the total length of the gel (mm) as measured from the bottom of a well.

6. Compare the migration of the α-galactosidase enzyme to the migration of the protein standards determined previously by Coomassie staining.

Visualization by Fluorescence

The third section of the gel, section "C," will be analyzed by a blotting process (i.e., a zymogram) using fluorescent detection. The assay is based on a fluorescent molecule 4-methylumbelliferyl linked to galactose. The detection occurs when the α-galactosidase cleaves 4-methylumbelliferyl-α-D-galactoside liberating the fluorescent moiety which is detected by long wavelength UV light.

The gel can then be Coomassie stained as described earlier to demonstrate the selectivity of the fluorescent detection. An example of the gel is shown in Figure 6.9.

Materials

4-methylumbelliferyl-α-D-galactoside (Sigma M7633)

0.25 M sodium acetate, pH 4.5

Whatman filter paper

UV transilluminator

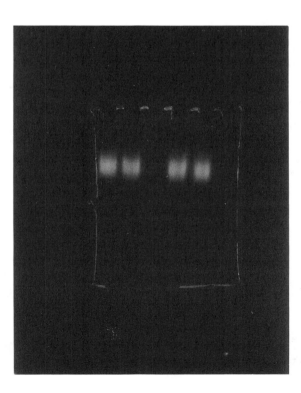

Figure 6.9 Fluorescent detection of α-galactosidase.

Method

1. A solution of 4-methylumbelliferyl-α-D-galactoside is prepared by dissolving 10 mg in 20 ml of 0.25 M sodium acetate buffer, pH 4.5. 4-methylumbelliferyl-α-D-galactoside is expensive and should be used both carefully and sparingly.

2. A piece of Whatman filter paper is cut to approximately the size of the gel section "C." The filter paper is saturated with the 4-methyl-umbelliferyl-α-D-galactoside substrate and is layered onto the gel section (which is left on the glass plate). The filter paper is pressed onto the gel by using a 13 × 100 mm test tube as a rolling pin.

3. The gel is incubated for approximately 5 minutes, and then the gel/glass plate is placed onto a UV transilluminator and analyzed by long wave UV radiation. The presence of α-galactosidase activity is indicated by an intense band of fluorescence on the gel. If a Polaroid camera is available, a photograph of the fluorescent gel should be taken.

4. The migration distance of the fluorescent band is measured (in mm) and divided by the length of the gel (in mm). Compare these numbers to the migration measured by the α-galactosidase assay technique using the gel slices.

5. Note that the lane containing the standard proteins does not fluoresce because the assay is specific for α-galactosidase.

6. The gel may now be Coomassie stained as described previously, and compared to the fluorescent detection.

STUDY QUESTIONS

1. Based on the structure of α-galactosidase, what is the predicted pattern that would be obtained if the enzyme were analyzed by SDS-PAGE?

2. What would be the effect of increasing acrylamide concentration on the electrophoresis of α-galactosidase?

3. Suggest a reason why the band corresponding to α-galactosidase in the native polyacrylamide gel is more diffuse than some of the protein standards. (Hint: α-galactosidase is a complex protein.)

4. What chemical procedure could be done to improve the sharpness of the bands corresponding to α-galactosidase in the native gel?

FURTHER READINGS

Janson J, Ryden L, eds. (1989): *Protein Purification, Principles, High Resolution Methods and Applications.* New York: VCH Publishers

Andrews A (1986): *Electrophoresis: Theory, Techniques, and Biochemical and Clinical Applications,* 2nd Ed. New York: Oxford University Press

Hames B, Rickwood D, eds. (1981): *Gel Electrophoresis: A Practical Approach.* Oxford: IRI Press

REFERENCES

Lazo PS, Ochoa AG, Jascón S (1978): α-Galactosidase (melibiase) from *Saccharomyces carlsbergensis*: Structural and kinetic properties. *Arch Biochem Biophys* 191:316–324

Chapter 7

Designing a Cloning Scheme

7.1 OVERVIEW

Cloning a gene is similar to finding a needle in a hay stack. A gene, even in the simplest of organisms, represents only a tiny fraction of the DNA in a genome (all the genetic information in the cell). A typical scheme for cloning involves removing the genomic DNA from a donor, breaking the DNA into many thousands of small pieces, and then searching those pieces for the desired gene. The search is accomplished by using molecular probes that adhere specifically to the targeted gene.

Strategies for cloning DNA differ markedly from those used for isolating and purifying proteins. Cloning relies on using nucleic acid probes to search for specific DNA or on introducing genome fragments into new hosts in which the targeted gene is expressed. Probes are used to locate DNA while expressed genes are usually identified by phenotypic changes in the host. Cloning does not always start with DNA. Messenger RNA (mRNA) can be isolated and converted into complementary DNA (cDNA). As we will see, cloning from either DNA or RNA each has merit depending on the circumstances.

The first half of this manual focused on the production, purification, and analysis of α-galactosidase. The final product of the purification, which was isolated by native PAGE, would have been subsequently sequenced. The resulting amino acid sequence is the data necessary to link the protein to the gene that codes for α-galactosidase.

This chapter will provide an overview to the strategies and approaches used in cloning. A major strategy for cloning applies protein sequence data, and this is the strategy demonstrated in this manual. The laboratory exercise will emphasize the most important aspect of experimentation, namely planning and strategy.

7.2   BACKGROUND

Cloning requires an understanding of the biochemistry of DNA and the higher order structure of the gene. The following section is an overview and not intended as an in-depth analysis of DNA and gene structure.

Biology of DNA

Nucleic acids are linear polymers of nucleotides connected by phosphodiester bonds. Nucleotides are composed of three parts: a phosphate group, a pentose (five carbon sugar), and a base (i.e., a heterocyclic organic base). In ribonucleic acid (RNA) the pentose is always ribose while in deoxyribonucleic acid (DNA) the pentose is deoxyribose, which is a ribose with a hydrogen at the 2' carbon rather than a hydroxyl (Figure 7.1).

There are two types of bases in nucleic acids, purines and pyrimidines. Pyrimidines contain one heterocyclic ring while purines are derivatives containing two fused rings. The rings are heterocyclic (contain one or more elements in addition to carbon), being composed of nitrogen and carbon. It is the nitrogen that gives these molecules their basic character, but none are ionized at neutral pH. The purines are adenine (A) and guanine (G), and the pyrimidines are cytosine (C), thymine (T), and uracil (U) (Figure 7.2). DNA contains the bases A, G, C, and T while RNA contains A, G, C, and U (i.e., the thymine found in DNA is replaced with uracil, a demethylated version of thymine).

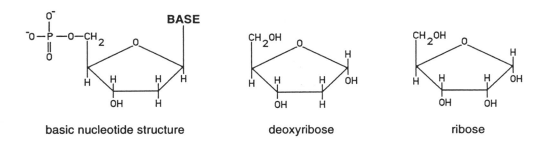

basic nucleotide structure   deoxyribose   ribose

Figure 7.1   Basic nucleic acid structures.

Like all organic molecules, the atoms of nucleotides are numbered. The bases are numbered with standard Arabic numerals (Figure 7.3) while the carbons of the pentose are designated with a prime so as to differentiate them from the atoms in the bases. The base and sugar are connected through the 1' carbon of the sugar and the 9 nitrogen of a purine or the 1 nitrogen of a pyrimidine. The 5' OH group of the sugar is linked by an ester bond to a phosphate group. The presence of a phosphate group, which is ionized at physiologic pH in nucleic acids, is what gives them their acidic character and net negative charge.

The Bases of DNA and RNA

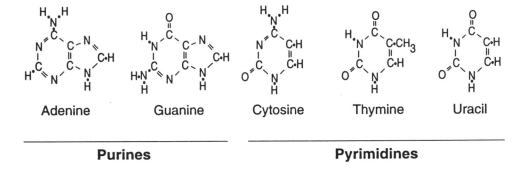

Adenine   Guanine   Cytosine   Thymine   Uracil

Purines   **Pyrimidines**

Figure 7.2   The purines and pyrimidines.

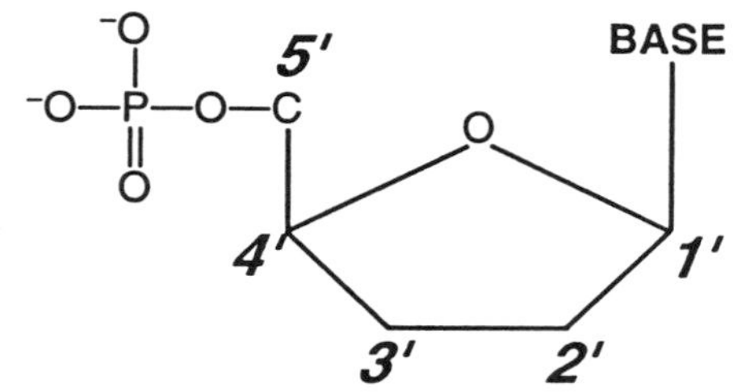

Figure 7.3   Purine, pyrimidine, and pentose numbering. Note the use of primes.

Nucleic acids may be formed both in vivo and in vitro by the 5' to 3' (i.e., 5'→3') polymerization of nucleotides via a condensation reaction. The 3' hydroxyl group of one nucleotide forms an ester bond with the 5' phosphate of another nucleotide. This leaves a free phosphate group at the 5' end and a free hydroxyl group at the growing 3' end (Figure 7.4). This reaction is catalyzed by DNA polymerase, but it can also be formed synthetically in a 3'→5' direction using phosphoamidite chemistry.

DNA usually exists within the cell as a double helix, the exception being single-stranded DNA viruses. The double helix has two right-handed helical polynucleotide chains which are coiled around a common axis. The chains are antiparallel with one chain running 5'→3' and the other running 3'→5' (Figure 7.5). The purine and pyrimidine bases are on the inside of the helix, with the planes of the bases perpendicular to the helix axis, whereas the phosphate and deoxyribose units are on the outside. The planes of the pentoses are nearly at right angles to those of the bases.

The two chains are held together by hydrogen bonds between pairs of bases (Figure 7.6). Adenine is always paired with thymine, while guanine is always paired with cytosine. Hydrophobic interactions between bases stacked above one another also contribute to the stability of the molecule. The negative charge of the phosphates results in repulsion of the backbones, a factor offset by base pairing.

The sequence of bases along a polynucleotide chain is not restricted. Nucleotide sequences form genes with two basic units: (1) the coding region comprised of codons, three base units that code for specific amino acids; and (2) regulatory sequences that define specific binding sites for various proteins or affect the local DNA structure in some way. These

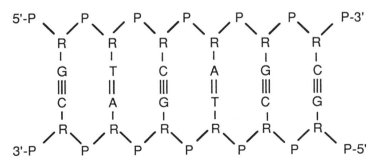

Figure 7.4 The basic structure of a DNA polymer.

informative and regulatory regions together constitute a gene, and combined are typically hundreds to thousands of nucleotides in length.

A gene (cistron) can generally be defined as a segment of DNA involved in the production of a polypeptide chain, and it includes a regulatory region (i.e., promoter), followed by a coding region, and bordered by a termination region. The promoter regulates the expres-

```
5'-P           P       P       P       P       P       P-3'
     \   /  \   /  \   /  \   /  \   /  \   /
      R       R       R       R       R       R
      |       |       |       |       |       |
      G       T       C       A       G       C
      |||     ||      |||     ||      |||     |||
      C       A       G       T       C       G
      |       |       |       |       |       |
      R       R       R       R       R       R
     /   \  /   \  /   \  /   \  /   \  /   \
3'-P       P       P       P       P       P-5'
```

Figure 7.5 Complementation of antiparallel strands of DNA.

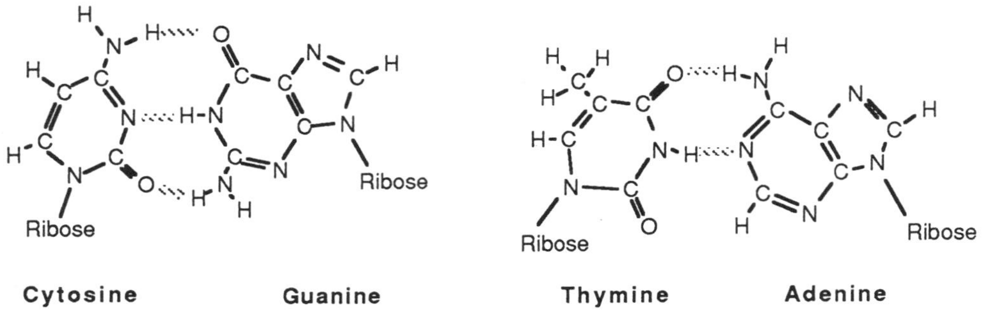

Figure 7.6   Hydrogen bonding between nucleotide bases.

sion of the gene, i.e., how much of the gene product is produced. The coding region dictates the linear sequence of amino acids in the gene product (protein), the exception to this being genes that code for RNA molecules which are used for a variety of cellular functions.

Eukaryotic coding regions are complicated due to the possible presence of noncoding regions *within* the gene. These noncoding regions are called **introns** while the coding regions are **exons** (Figure 7.7). RNA transcribed from such a gene is processed, and the introns are removed (spliced out) prior to translation. Introns can comprise a substantial portion of eukaryotic genes.

Transcription is the process by which an RNA message (i.e., mRNA) is synthesized from the gene. Once initiated, transcription proceeds from

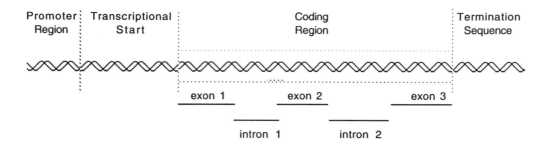

Figure 7.7   Structure of a eukaryotic gene.

the promoter and continues through the coding region. For conservation purposes, transcription is terminated shortly after the coding region has been copied. Following the end of the coding region can be found a region containing termination sequences, such as an inverted repeats. Terminators allow the RNA to form a secondary structure that destabilizes and halts the transcription process.

Cloning Strategies

Cloning a gene is an involved process relying not only on molecular techniques, but also on classical genetics, protein biochemistry, and microbiology. The initial stages of cloning are straightforward and involve: (1) isolation of DNA from the source organism; (2) fragmentation of the DNA into relatively small pieces; (3) combining the fragments with a vector (a DNA molecule which replicates in a host) to form a recombinant DNA molecule; (4) introducing the recombinant DNA into a host; and (5) screening hosts for a specific fragment of DNA.

Often millions of recombinant molecules must be examined to identify one specific clone. Consequently strategies that simplify the identification process need to be adopted.

One strategy for identifying a recombinant molecule relies on the ability of the cloned gene to provide a missing function, or to complement a mutation, in a new host (Table 7.1). Complementation is a straight forward approach to cloning; however, its application is limited. It is not

Table 7.1   Examples of Genes Cloned by Complementation

Gene	Source	Function	Host	Selection	Comments
TetA	S. aureus	Tetracycline resistance	E. coli	Tetracycline	Tetracycline is an inhibitor of protein synthesis
LEU2	S. cerevisiae	Leucine biosynthesis	E. coli	Leucine minus medium	E. coli is leu⁻
MEL1	S. carlsbergensis	Melibiose hydrolysis	S. cerevisiae	Growth on melibiose	S. cerevisiae is mel⁰
NEO	Transposon	Aminoglycosid ase	Yeast	Growth on G418 (antibiotic)	Antibiotic G418 is destroyed by aminoglycosidases

common for genes from one organism to function properly in a new host unless the two organisms are similar. This is not because the protein (i.e., gene product) is incapable of biological activity, but rather because the regulatory mechanisms for transcription are not recognized. It is much like putting an 1.4 Mb computer disk into an 800K disk drive in that the data is there, but the drive cannot recognize the format. Two highly divergent organisms are normally incapable of recognizing each other's genes.

More complex, but very similar to cloning by complementation, is cloning using complementary DNA (cDNA) and expression vectors. cDNA is not a normal cellular product but is synthesized in vitro from messenger RNA (mRNA). In more complex organisms, such as the higher eukaryotes (e.g., animals and vascular plants), the genomes are too large to readily manipulate for cloning. As such, the mRNA can be harvested from cells and converted into cDNA by the enzyme reverse transcriptase. Reverse transcriptase is an enzyme encoded and found in retroviruses (also called RNA tumor viruses). Retroviruses have RNA genomes and, upon infecting their hosts, convert the RNA into cDNA during infection.

cDNA represents only the region of a gene that acted as a template during mRNA synthesis. cDNA does not include the promoter and terminator regions of the original gene. Consequently, if cDNA is inserted into a new host, it would not express since no regulatory regions are available. To overcome this, cDNA is often cloned into expression vectors, i.e., plasmids or viruses that have promoters/transcriptional terminators lacking the normally intervening coding region. cDNAs can be inserted between and regulated from the promoter and terminator. Since the new recombinant molecule is constructed by combining regulatory and coding regions from different sources, such constructs are called heterologous genes. Heterologous cDNAs can be used to complement analogous mutations in many hosts. For instance, the human *CDC*28 cDNA (a gene fundamental to the initiation of DNA synthesis in dividing cells) has been expressed from a yeast promoter and then used to complement a *cdc*28 mutation in yeast.

Most genes are not cloned by complementation, but by searching a library with probes that hybridize (adhere) to the targeted gene. A library is a pool of either random genomic fragments (i.e., pieces of the cells' collective DNA) or cDNAs coupled to a vector. Vectors are DNA molecules that are used to carry the pieces of a library in host cells. Genomic libraries are constructed by enzymatically fragmenting all the available DNA from a donor, followed by linking those fragments to a vector. A cDNA library involves synthesizing cDNA (see above) and linking those molecules to a vector.

Once libraries are introduced into host cells, the cells and their recombinant DNA replicate. Each cell hosts a different library fragment or cDNA. These hosts can be individually cultured to yield a population of genetically identical cells. The term clone is often used to describe these cells or the specific recombinant DNA molecule inside the host. Specific DNA clones are identified by searching a heterogeneous population of hosts (usually in the form of colonies on a petri plate) with probes homologous to the targeted DNA molecule. Cells that bind the probe are cultured, and their DNA is isolated and further characterized.

Identifying Clones with Probes

Probes are specific, homologous nucleic acids, which are made either naturally or synthetically. Natural probes are derived from either cDNA or genomic DNA while a synthetic probe is an oligonucleotide (short DNA sequence of less than 100 bases) produced on an DNA synthesizer.

Oligonucleotide probes can be derived from the amino acid sequence of a protein which is a reflection of the nucleotide sequence of the corresponding gene. Since protein production involves the synthesis (transcription) of mRNA from a DNA template, and subsequent translation of the mRNA into a chain of amino acids, the amino acid sequence is representative of the DNA sequence. This relationship is the basis of the genetic code (Figure 7.8). The three nucleotides of the codon, i.e., the code for an amino acid, are arranged to represent the 20 amino acids and the stop codons, i.e., translation termination. However, the four nucleotide bases (i.e., A, C, G, and T) can be rearranged into 64 different combinations, or codons (Figure 7.8). As such, on average each amino acid is represented by more than one codon. This leeway in the coding for amino acids is termed degeneracy.

The amino acid sequence of a protein, as determined by protein sequence analysis (see Chapter 6), can be used to design a nucleic acid molecule that is representative of its corresponding gene. By reverse translating, i.e., going backwards from amino acids to codons, an amino acid sequence can be used to predict the corresponding nucleotide sequence.

The Genetic Code

Second Codon Base

		U	C	A	G	
	U	UUU - Phe UUC - Phe UUA - Leu UUG - Leu	UCU - Ser UCC - Ser UCA - Ser UCG - Ser	UAU - Tyr UAC - Tyr UAA - Stop UAG - Stop	UGU - Cys UGC - Cys UGA - Stop UGG - Trp	U C A G
First Codon Base	C	CUU - Leu CUC - Leu CUA - Leu CUG - Leu	CCU - Pro CCC - Pro CCA - Pro CCG - Pro	CAU - His CAC - His CAA - Gln CAG - Gln	CGU - Arg CGC - Arg CGA - Arg CGG - Arg	U C A G
	A	AUU - Ile AUC - Ile AUA - Ile AUG - Met	ACU - Thr ACC - Thr ACA - Thr ACG - Thr	AAU - Asn AAC - Asn AAA - Lys AAG - Lys	AGU - Ser AGC - Ser AGA - Arg AGG - Arg	U C A G
	G	GUU - Val GUC - Val GUA - Val GUG - Val	GCU - Ala GCC - Ala GCA - Ala GCG - Ala	GAU - Asp GAC - Asp GAA - Glu GAG - Glu	GGU - Gly GGC - Gly GGA - Gly GGG - Gly	U C A G

Third Codon Base

Figure 7.8 The 64 codons of the genetic code and their associated message. The code is presented as RNA; thus U is substituted for T.

However the degeneracy of the genetic code only allows for an estimate of the sequence and not an exact determination. Since all but two of the amino acids are represented by more than one codon (the exceptions being methionine [ATG] and tryptophan [TGG]), the construction of a nucleotide sequence is, at best, a guess. When synthesizing an oligonucleotide on a DNA synthesizer, allowances can be made for degeneracy by making mixed oligonucleotides, i.e., making a pool of oligonucleotides representing every possibility of a degenerate sequence (Figure 7.9).

A natural probe useful in genomic library screening is cDNA. Due to the enormous genomes of higher eukaryotes, cDNA is often synthesized prior to the cloning of the genomic equivalent. Not surprisingly cDNA is the perfect probe for screening a library since it is both specific and long. The presence of introns in genomic DNA does not alter the effectiveness of the cDNA probe since exons represented in the cDNA are

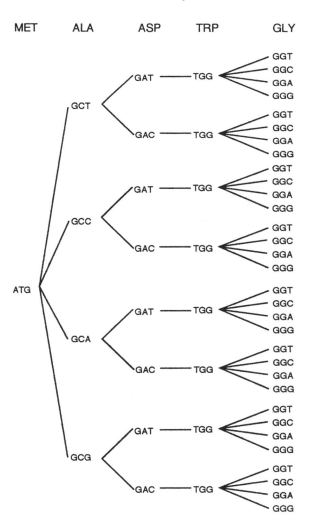

Figure 7.9 Degenerate oligonucleotides designed from an amino acid sequence. Schematic of possible nucleotide sequences which could code for the following pentapeptide.

normally several hundred bases in length, thus affording ample specificity for hybridization.

Interspecies genomic clones or interspecies cDNAs can also be used to probe a library. The basic criterion for the probe is that it must share DNA homology with the target. Often genes of similar species are used as probes with the assumption that genes with similar functions, are to some degree, evolutionarily conserved. The question of what defines a similar species in genetic terms is difficult to precisely define. For instance, two divergent species possessing analogous enzyme activities may have little homology between their DNAs. More closely related species, such as mice and rats, are often considered genetically similar when genes have 50% or greater DNA homology. For a cloned gene to act as a probe, it simply binds to the comparable gene in the library. Conserved regions between the two genes, i.e., regions coding for active sites or critical structural regions, usually have sufficient homology to hybridize.

Probe Labelling Options

As noted above, probes must contain labels that can be detected after hybridization to the target sequence. The means of labelling will depend on whether the probe is an oligonucleotide or cloned DNA (cDNA or genomic). Generally, labels are either radioactive (hot) nucleotides or specifically modified nucleotides in which a nonradioactive marker is attached (cold). The options for incorporating these labelled nucleotides into a probe are as follows:

- Tailing—Terminal deoxytransferase (TdT) from calf thymus adds nucleotides onto the 3' end of single-stranded, blunt-ended or 3' overhangs (i.e., double-stranded DNA in which a 3' end protrudes from the end of a double helix). By mixing labelled nucleotides, probe DNA, and TdT, labelled nucleotides can be incorporated into tails without affecting the specificity of the original sequence (Figure 7.10).

- 5' Labelling—Oligonucleotides are often labelled at the 5' end during synthesis or postsynthetically. During synthesis, cold labels can be added to the nucleotide during the final step of synthesis. A common cold label is biotin, which can be

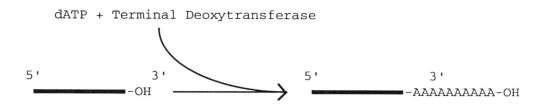

dATP + Terminal Deoxytransferase

5' ———————————3' 5' ———————————3'
————————————-OH ————→ ——————————-AAAAAAAAAA-OH

Figure 7.10   Addition of nucleotides to DNA using terminal transferase.

indirectly detected enzymatically. Alternatively, 5' phosphates of DNA can be removed with alkaline phosphatase, yielding a 5' hydroxyl. Hot dATP (γ-^{32}P) can be used to replace the phosphate on the DNA via polynucleotide kinase, which will transfer the γ phosphate from the dATP to the DNA.

▪ Random Priming—Cloned genes and cDNA easily can be labelled through random priming. This technique involves denaturing the DNA (by boiling) to single-strands followed by the annealing of random hexamers to the exposed templates. Random hexamers are simple six base oligonucleotides, all of which are random in their nucleotide sequence. These random hexamers anneal to the template and act as primers for DNA synthesis. By adding a polymerase, all four deoxynucleotide triphosphates, and a labelled nucleotide analog (e.g., digoxigenin-11-dUTP which is recognized as dTTP), the gaps between the annealed primers are filled in by the polymerase. The digoxigenin, in this nucleotide analog, can be detected by an enzyme immunoassay, and thus is a cold label. The newly synthesized DNA will contain a number of these labelled bases.

▪ In Vitro Transcription—A highly specific technique used for synthesizing probes involves the in vitro synthesis of mRNA, i.e., in vitro transcription. Although not commonly used for library screening, mRNA probes are readily used for Northern blotting, a technique used to detect specific mRNA from cells. The synthesis usually involves cloning a cDNA into a specifically designed expression vector in which the cDNA is flanked by phage promoters (i.e., bacterial virus). When purified phage RNA polymerase and nucleotide triphosphates are mixed with the expression vector, mRNA is synthesized from the cDNA strand. For the probe to be homologous to the mRNA synthesized in vivo, the probe must be antisense or synthesized off the wrong DNA template. The antisense mRNA is homologous to and will hybridize to the mRNA. This antisense probe can easily be synthesized in vitro by controlling which RNA polymerase is used for the transcription (Figure 7.11). RNA probes are highly specific and are useful for Northern blotting, in situ hybridizations, and RNase protection assays.

▪ Nick Translation—An important enzyme in *E. coli* DNA replication is DNA polymerase I (pol I). This enzyme has three activities, which are a 5'→3' DNA polymerase activity, a 3'→5' proofreading function (exonuclease), and a 5'→3' exonuclease activity. The proofreading activity ensures that the correct nucleotide is being incorporated into a growing DNA chain while the exonuclease activity degrades one strand of a helix in a 5'→3' direction from nicks in the DNA. The polymerase and 5'→3' exonuclease activities can be used for the synthesis of probes. By introducing random nicks into DNA backbone with low concentrations of endonucleases (an enzyme which breaks the bond between a ribose and a phosphate), a template is created for the polymerase and exonuclease activities of pol I. The addition of deoxynucleotides triphosphates, a labelled nucleotide triphosphate, such as α-^{32}P labelled ATP, and DNA polymerase I, will result in the degradation of a DNA strand and its replacement with a new strand that contains a label. The nick in the DNA which is critical for the probe synthesis, actually moves 5'→3' as the polymerase degrades and lays down a new template; thus the probe synthesis is called nick translation. This early approach to labelling DNA generally has been replaced by random priming.

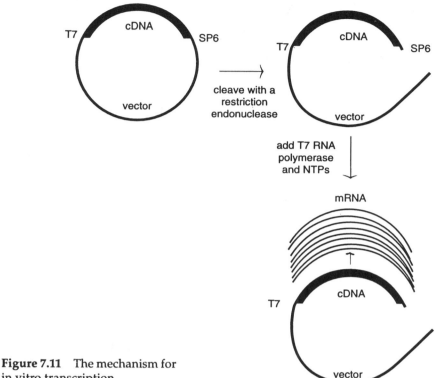

Figure 7.11 The mechanism for in vitro transcription.

7.3 EXPERIMENTAL DESIGN AND PROCEDURES

There are many approaches to cloning a gene, and this manual will focus on one, namely, the purification and analysis of a protein to produce a synthetic oligonucleotide probe for cloning. The first six chapters of this manual focused on the protein, while the remaining chapters will concentrate on the cloning and characterization of the DNA. The experimental exercises of this chapter will be to design cloning experiments, prepare media, and culture yeast and *E. coli*. The DNA from these cultured cells will be isolated subsequently in Chapter 8.

Designing a Cloning Scheme

Planning, designing, and implementing a research project is a very important, but often overlooked aspect of science education. This type of responsibility usually doesn't arise until graduate study, and even then the student may only have minor input. In industry, junior scientists will often work mechanically on a project without knowing the objective of the experiment or even their own role in the research scheme. However, it is becoming increasingly necessary for all the individuals of a research team to take responsibility for the research process.

The objective of this exercise is for you to take the responsibility of designing a research scheme for cloning the yeast gene that encodes α-galactosidase. (The actual gene is called *MEL1*.) The key to this task is not to get bogged down with details, but to propose a general scheme for this project. Simply base this scheme on the information presented. The details should not refer to specific techniques, but rather to information that will support the idea that your approach is feasible and will work. Consider the following questions during this task:

- What options do you have for identifying the *MEL1* gene from *Saccharomyces carlsbergensis*?

- What is your approach for cloning this gene? Demonstrate the approach by designing a cloning flow chart analogous to the protein purification scheme (see Figure 2.7).

- How can you support your cloning strategy? Why is your method better than alternative methods?

Designing a Probe—DNA from Protein

A common approach to identifying a clone is to design a homologous oligonucleotide probe based on the protein's amino acid sequence. These oligonucleotides are short, single-stranded, synthetic DNA or RNA molecules. The cloning scheme presented in this manual relies on purifying α-galactosidase from yeast and then sequencing the protein to produce data which can be used to make a probe. That probe will act like a magnet and specifically adhere to the *MEL1* gene. By reverse trans-

Table 7.2 The Number of Codons for Each Amino Acid

Number of Codons	Amino Acid
1	met, trp
2	asn, asp, cys, gln, glu, his, lys, phe, tyr
3	ile, Stop Codons
4	ala, gly, pro, thr, val
6	arg, leu, ser

lating the amino acid sequence, that is by preceding backwards from protein to nucleic acid, a DNA molecule can be designed that is complementary to the targeted gene. A good protein sequencing reaction will yield up to 50 N-terminal amino acids. It is from this N-terminal region that the probe must be designed.

In this exercise you will manually design an oligonucleotide probe from the *MEL1* N-terminal amino acid sequence. This sequence is derived from the native PAGE purified α-galactosidase isolated in Chapter 6. The sequence of the 50 N-terminal amino acids of α-galactosidase is as follows:

> met-phe-ala-phe-tyr-phe-leu-thr-ala-cys-ile-ser-leu-lys-gly-
> val-phe-gly-val-ser-pro-ser-tyr-asn-gly-leu-gly-leu-thr-pro-
> gln-met-gly-trp-asp-asn-trp-asn-thr-phe-ala-cys-asp-val-ser-
> glu-gln-leu-leu-leu

Designing a probe from the amino acid sequence requires identification of a continuous region that reverse translates into a minimum number of sequences (known as permutations). By selecting amino acid sequences predominantly consisting of Trp, Met, Phe, Tyr, His, Gln, Asn, Lys, Asp, Glu, and Cys, the variations or permutations in the oligonucleotide sequence can be minimized. Table 7.2 summarizes the number of codons per amino acid in the genetic code.

Your task is to locate a region within the amino acid sequence that has few permutations (i.e., variations) when reverse translated. The resulting oligonucleotide should be between 15 and 25 bases in length.

Longer probes are more desirable because they are more specific. Probes that are approximately 50% guanines and cytosines are also useful.

Hint: Try listing the amino acids on paper with the number of codons per amino acid underneath each amino acid. Multiplying the numbers of adjacent amino acids provides the number of permutations for that sequence. Look for a sequence with the least number of permutations.

Preparation of Media for Yeast and *E. coli*

The strains of *Saccharomyces carlsbergensis* and *Escherichia coli* that are used for the cloning portion of this experiment must be activated and cultured. The yeast will donate genomic DNA to the cloning scheme while the *E. coli* will provide the cloning vector. The yeast can be cultured in a standard medium, such as YPD, while *E. coli* must be cultured under selective pressure. The *E. coli* possesses a plasmid (a small autonomously replicating DNA molecule) which will serve as the vector. The plasmid contains a gene for ampicillin resistance which allows the bacteria to be grown in the presence of the antibiotic. Only *E. coli* that grow in ampicillin will contain the plasmid.

Materials

Tryptone
Sodium chloride
Yeast extract
Peptone
Glucose
Ampicillin, sodium salt
Agar
Sterile petri dishes
Flasks
Foam stoppers
Aluminum foil
Autoclave tape

Method

1. Four different media must be prepared for the cultivation of yeast and *E. coli*. For the yeast, YPD agar plates and YPD broth must be prepared while *E. coli* requires LB-ampicillin agar plates and LB-ampicillin broth.

2. Prepare 2 ml of a 25 mg/ml solution of ampicillin in water. Once the antibiotic is dissolved, it is filter sterilized using a syringe and 0.45 μm membrane syringe filter. The solution should be aliquoted (1 ml) and stored at –20°C.

3. Using techniques described in Chapter 1.3, prepare the following:

   YPD Broth—2% glucose, 2% peptone, 1% yeast extract. Prepare 50 ml and autoclave in a 250 to 500 ml flask.

   YPD Agar—The same recipe as YPD broth but with 2% agar. Prepare 100 ml, autoclave, and then pour five petri plates.

   LB Broth-Amp—1% Tryptone, 1% NaCl, 0.5% yeast extract. Prepare 20 ml and autoclave in a 100 ml flask. Add ampicillin (2 μl/ml) just before use.

   LB Agar-Amp—The same recipe as LB broth-amp but with 2% agar. Add ampicillin after cooling the agar to 55°C. Prepare 100 ml and pour five plates.

Once solidified, the plates can be stored at 4°C until use. It is best to invert the plates during storage (condensation will drip on the surface and increase the chance of contamination).

Streaking of Stored *S. carlsbergensis* and *E. coli*

From stock cultures, activate and streak both the yeast and *E. coli* onto YPD agar and LB-amp agar, respectively. For a review of streaking techniques, refer to Chapter 1.3.

Many options are available for choosing which yeast and *E. coli* should be used for the following series of experiments. For *Saccharomyces carlsbergensis*, as noted previously, any α-galactosidase positive strain is suitable for cloning. The strain which was used earlier for the production of α-galactosidase is the obvious selection.

The choice of *E. coli* strain is more difficult since it must contain a cloning vector as well as be a suitable host. Two good *E. coli* strains which are available for recombinant DNA research are DH5α and TOP10F' (Invitrogen). Either one of these strains that harbor any of the following plasmids would be suitable for the subsequent laboratory experiments: pUC18, pUC19, pUC118, pUC119, pUCBM20 (Boehringer Mannheim), pUCBM21 (Boehringer Mannheim), and pDNAII (Invitrogen).

Unfortunately, plasmids are not usually sold in *E. coli*, but rather as purified DNA. Once plasmids are purchased, they are typically introduced (transformed) into a desirable *E. coli*, cultured, and then stored at –20°C or –80°C. Unless the plasmid DNA is specially treated, it only needs to be purchased once, since after its introduction into *E. coli*, the bacteria can be cultured and the plasmid harvested as needed. If your plasmid is purified, introduce it back into a strain of *E. coli*, such as DH5α or TOP10F'. This is described in the transformation experiment in Chapter 10.3.

Inoculation and Culturing of Yeast and *E. coli*

Both the yeast and bacteria must be cultured to provide adequate numbers of cells for the harvesting of genomic and plasmid DNAs, respectively. The first step is to prepare actively growing cells (as above) and to isolate an individual colony. From this colony you will inoculate the broths as described in Chapter 1. Inoculate the yeast into the YPD broth and *E. coli* into LB broth. Ampicillin must be added to the LB broth (2 µl/ ml) before use. Incubate the cultures for 24 to 48 hrs as described in Chapter 1.

Materials

Inoculation loop
Bunsen burner or alcohol lamp
YPD broth

MORE...

LB broth

Ampicillin solution—25 mg/ml, frozen at –20°C.

Shaking incubator or equivalent

Method

1. Thaw the ampicillin solution. Aseptically add 2 µl ampicillin/ml of LB broth.

2. Using aseptic technique, inoculate the LB-ampicillin broth with a small quantity of cells from an isolated *E. coli* colony.

3. Similarly, inoculate the YPD broth with a small quantity of yeast colony.

4. Incubate the *E. coli* overnight at 37°C with shaking. The yeast should be cultured 24 to 48 hrs at 30°C with shaking. If only one incubator shaker is available, both organisms can be cultured at 30°C. If necessary, both organisms can be grown at room temperature. If no incubator shaker is available, then sterile magnetic stir bars can be added to the broths. The flasks can then be agitated on magnetic stir plates either in incubators or at room temperature.

5. Both the yeast and *E. coli* cultures can be frozen until use, though it is best to use them fresh.

STUDY QUESTIONS

1. What strategies can you devise to clone the *MEL1* gene by complementation?

2. Investigate the mechanism by which oligonucleotide probes are synthesized.

3. What are the advantages and disadvantages of cDNA cloning compared to genomic cloning?

FURTHER READINGS

Lathe R (1985): Synthetic oligonucleotide probes deduced from amino acid sequence data: Theoretical and practical applications. *J Mol Biol* 183:1–12

Suggs SV, Wallace RB, Hirose T, Kawashima EH, Itakura K (1981): Use of synthetic oligonucleotides as hybridization probes: Isolation of cloned cDNA sequences for human β_2 microglobulin. *Proc Natl Acad Sci USA* 78:6613–6617

Watson J, Tooze J, Kurtz D (1983): *Recombinant DNA: A Short Course.* New York: Scientific American Books

8

Isolation and Preparation of Nucleic Acids

8.1 OVERVIEW

Before anything can be constructed, it is first necessary to gather the building materials, and, therefore, harvesting genomic DNA is a prerequisite for cloning. Normally two types of DNA are required for cloning, namely, the source DNA containing the targeted gene (that which is to be cloned), and the vector (a DNA molecule that carries the target). The source or genomic DNA can be from any organism or DNA virus. The vector, on the other hand, is a specially designed DNA molecule derived from either a bacteriophage, a plasmid, or some combination of both. As we shall see, the vector will serve as a carrier for the genomic DNA fragments.

Methods for isolating nucleic acids depend on its source. It generally involves collecting cells or tissues, disrupting the cells with enzymes and/or detergents, separating the nucleic acids from other biomolecules and cellular debris, and finally concentrating and/or drying the nucleic acid. The complexity of isolating DNA differs depending on the source. Viral DNA simply requires stripping away the protein coat with an organic solvent (e.g., phenol and chloroform), while plant cells first require removal of the thick cell wall by physically pulverizing the cells (Table 8.1).

Several different schemes can be used for the cloning of the *MEL1* gene encoding α-galactosidase. If the complete gene is to be cloned, then the genomic DNA must be isolated. This is in contrast to isolating yeast mRNA and synthesizing and cloning cDNA. This chapter will examine the means of isolating nucleic acids, measuring the yield, and techniques for their storage. The experimental focus will be on the isolation of genomic DNA from the yeast *Saccharomyces carlsbergensis* (source) and plasmid DNA (vector) from *Escherichia coli*. The yeast is the source of the *MEL1* gene that encodes α-galactosidase.

8.2   BACKGROUND

The eukaryotic cell may have several different types of DNA. Each cell contains multiple chromosomes, but it may also harbor mitochondrial, chloroplast, and plasmid DNAs. Regardless of the particular source, all the species of DNA within a cell are referred to as the genome. The isolation and cloning of this DNA is called genomic cloning, which is in sharp contrast to cDNA cloning. Once genomic DNA is extracted from an organism it can be manipulated with a standard set of techniques. However, depending on the organism used, the method of harvesting the genomic DNA can vary extensively (Table 8.1).

For microorganisms, the cells must be cultured in a rich medium (e.g., glucose, yeast extract, and amino acids) until dense and then harvested by centrifugation. In an isotonic solution, the microbial cell wall is removed with an enzyme (e.g., chicken lysozyme for bacteria), followed by the emulsification of the membrane with a mild detergent (e.g., Triton X-100). Within the cell the chromosome is tightly packaged and coiled so bursting the cell with a harsh detergent, such as SDS, will cause the chromosome to rapidly expand and break (called shearing). Mild detergents allow the cell to slowly fall apart thus preventing chromosomal shearing.

Harvesting DNA from animal cell culture is relatively easy since tissue culture lacks cell walls. Generally a mild detergent such as NP-40 is added to cells along with the protease Proteinase K. NP-40 lyses the cells while

Table 8.1   Approaches to Isolating Genomic DNA

Organism	Treatment
Phage	Inoculate susceptible cells, collect lysate, extract protein, precipitate DNA.
Bacteria	Culture cells, treat isotonically with lysozyme, add mild detergent (e.g., Triton X-100), purify DNA on a CsCl gradient.
Yeast	Grow cells, treat isotonically with glusulase, lyse cells with mild detergent (i.e., Triton X-100, sarkosyl), purify DNA on a CsCl gradient.
Mold	Culture and harvest mycelia, treat isotonically with novozyme, lyse with detergent, purify DNA with a CsCl gradient.
Plant	Freeze leaves in liquid nitrogen, grind cells, lyse with SDS, purify DNA via CsCl gradient.
Tissue Culture	Using a monolayer, lyse cells by treatment with SDS and Proteinase K at alkaline pH (9), spool DNA onto glass rod, and dissolve in TE buffer.
Organ	Freeze tissue and grind in liquid nitrogen, lyse cells with SDS, remove protein with Proteinase K and phenol, spool DNA onto glass rod, and dissolve in TE buffer.

Proteinase K degrades interfering proteins such as histones and nucleases. Degrading histones liberates the DNA from its package, and eliminating nucleases protects the DNA from undue degradation.

Plasmid isolation differs markedly from the techniques used for genomic DNA due to the contrasting natures of the molecules. Culturing bacteria for a plasmid isolation involves growing the organism under selective pressure, i.e., in conditions which establish a vital role for the plasmid encoded genes. With *E. coli,* many plasmids rely on the β-lactamase gene to provide resistance to the antibiotic ampicillin (ampicillin interferes with the synthesis of the peptidoglycan in the cell wall). β-Lactamase is an enzyme (encoded by Am^R) that cleaves and destroys ampicillin. Culturing *E. coli* in a nutrient broth with ampicillin (50 μg/ml) applies a selective pressure so that the plasmid is maintained by the cell. In some organisms, such as *S. cerevisiae*, plasmids are maintained through nutritional pressures. In this situation, the plasmid contains a gene complementing a host mutation. The plasmid is maintained by growing the organism in a minimal medium lacking the nutrient that will be provided by gene(s) encoded within the plasmid. In both instances, cells that lose the plasmid either become stagnant or die.

Both genomic and plasmid DNA preparations normally contain contaminating proteins which often need to be removed. One common method of removing contaminating protein is by phenol:chloroform extraction. Phenol and chloroform are organic solvents that denature and remove the proteins from aqueous solutions. Both solvents separate from, and are heavier than, water so that mixing of a DNA solution with phenol:chloroform results in a lower organic phase containing the protein fraction. Normally these extractions make use of a 25:24:1 mixture of phenol, chloroform, and isoamyl alcohol. The isoamyl alcohol helps in the separation of the organic and aqueous phases. Solutions with extremely high concentrations of protein have a white precipitate at the interface of the phases, which disappears after repeated extractions. Residual phenol in the aqueous phase can interfere with downstream manipulations of the DNA; thus the DNA is often precipitated by the addition of salt and ethanol (see Chapter 8.3).

Frequently DNA species need to be purified from one another, e.g., when chromosomal fragments contaminate a plasmid preparation. A powerful method of separating different species of DNA, while also removing other impurities, makes use of ultracentrifugation. When a nucleic acid solution is mixed with cesium chloride and centrifuged at an extremely high speed (e.g., 42,000–60,000 rpm), the cesium ions form a density gradient within the centrifuge tube. In this gradient, the nucleic acids migrate to a density equal to their own. When mixed with the fluorescent dye ethidium bromide (also toxic and carcinogenic), protein, RNA, plasmid DNA, and chromosomal DNA are separated from each other based on their densities (Figure 8.1). Within a cesium gradient, protein, being the least dense, is usually on top, followed by the chromosomal DNA, then plasmid DNA, and finally by RNA. The separation of the plasmid/chromosomal DNAs is based on the supercoiled nature of plasmid DNA. Nonsupercoiled DNA, known as relaxed DNA, has a lower density in ethidium bromide than does supercoiled DNA, which is the normal conformation of a plasmid. If a plasmid is nicked or sheared, it acquires the density of relaxed DNA. As such, relaxed DNA often contains nonsupercoiled plasmid in addition to chromosomal DNA.

Alternative methods are available for purifying DNA, such as ion exchange chromatography. Under basic conditions (e.g., pH 8), DNA

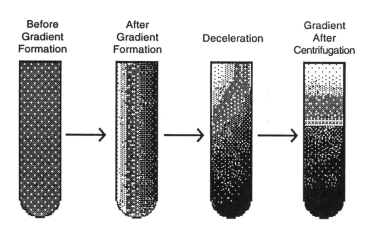

Before Gradient Formation	After Gradient Formation	Deceleration	Gradient After Centrifugation

Figure 8.1 Separation of nucleic acids by ultra-centrifugation. This representation of a vertical CsCl gradient starts with a homogeneous solution which forms a very narrow gradient. Upon deceleration, the gradient reorients so as to be horizontal.

has a negative charge due to the ionized phosphates in the phosphoribose backbone. By employing an anion exchange column, DNA can be adsorbed to the column and then eluted with an increasing salt gradient in the same manner as protein purification. Purification occurs as different species of DNA elute at different salt concentrations. Specialized resins and columns (such as the Qiagen™ column) are available for ion exchange purification. Ion exchange can also be used in batch mode to isolate DNA from protein, and application of these techniques often results in DNA of high purity.

Measuring Nucleic Acid Concentrations

The concentration of nucleic acids in solution can be determined by UV spectroscopy, either qualitatively or quantitatively, since DNA absorbs light strongly at 260 nm. Qualitatively, double and single-stranded DNA and RNA have been assigned average absorbance values. At 260 nm, 1 OD unit has been estimated to contain the following:

double-stranded DNA	50 μg/ml,
single-stranded DNA	33 μg/ml,
single-stranded RNA	40 μg/ml.

Pure solutions of any one of the above nucleic acids could easily be assayed for concentration. The difficulty arises when the preparation has more than one type of nucleic acid. For instance, plasmids isolated from *E. coli* typically include RNA unless an RNase is added during the protocol to degrade the contaminant. More problematic is the presence of contaminating protein. Since protein also absorbs in the 260 nm range, nucleic acid solutions tainted with protein yield false concentration values. Prior to determining the concentration, nucleic acids should be phenol/chloroform extracted to remove contaminating proteins.

Quantitatively, the concentration of nucleic acids can be measured using Beer's Law, which states:

$$A = \varepsilon bc,$$

where "A" is absorbance, "b" is the light path (cm), "c" is the concentration of the solute (M), and ε is the extinction coefficient. Each molecular species of DNA has its own extinction coefficient which is dependent upon its length and base composition. The extinction coefficient is determined by summing the individual ε for each base. The extinction coefficients for the individual bases are:

adenine	15200
cytosine	7050
guanine	12010
thymine	8400

Beer's Law is only practical if the DNA molecule has been highly purified and has a known sequence, such as with a synthesized oligonucleotide. Beer's Law can be applied to larger DNA molecules, but the calculation of ε is usually cumbersome. For larger DNA molecules, the qualitative measurement is more practical.

Once DNA has been isolated, extracted, and quantitated, it needs to be carefully stored and preserved. DNA is a very stable molecule, thus storing DNA can be accomplished in many ways. The most stable method is to dry the DNA in a lyophilizer or a SpeedVac®. SpeedVacs are essentially centrifuges connected to a vacuum pump. Typically, a solution of DNA is centrifuged while a vacuum is applied and the resulting dry pellet can be stored frozen for years. Simply freezing DNA solutions is also acceptable, but removing the water (lyophilization) prevents hydrolysis of the DNA over time. When DNA is in solution, it is advisable to add EDTA (1 mM) which chelates or binds divalent cations, such as Mg^{+2}, which is needed by DNases (DNA degrading enzymes).

8.3   EXPERIMENTAL DESIGNS AND PROCDURES

The objective of the following laboratory experiments is to harvest genomic DNA from *S. carlsbergensis* and vector (plasmid) DNA from *E. coli* for preparation in cloning. The gene we plan to clone is the yeast *MEL1* which encodes α-galactosidase. As noted in Chapter 7, the source DNA must be inserted or linked into a vector, e.g., a plasmid from *E. coli*. These experiments will accomplish the isolation of genomic DNA, plasmid DNA, removal of protein from the nucleic acids, determination of their concentrations, and finally storage for future use.

Yeast Genomic DNA Isolation

Isolating genomic DNA from yeast involves culturing the microbe, harvesting the cells, enzymatically removing the cell wall, lysing the protoplasts, and finally separating the DNA from the other cell debris. The initial steps of this protocol are analogous to yeast protoplast formation performed in Chapter 3. However DNA isolation requires the additional steps of lysing the cells and then separating the DNA from contaminating biomolecules.

Materials

Yeast culture—prepared previously

Spectrophotometer with cuvettes

50 mM EDTA, pH 8—ice cold

50 mM Tris, pH 9.5, 2% 2-mercaptoethanol

1.2 M sorbitol, 50 mM Tris, pH 7.5

Lyticase solution—500 U/ml in 50 mM Tris, pH 7.5

MORE...

10% SDS—used for checking protoplast formation

Lysis buffer—100 mM Tris, pH 7.5, 100 mM EDTA, 150 mM
    NaCl, 50 µg/ml RNase A

Lysis buffer with 2% SDS

95% Ethanol—stored at –20°C

TE Buffer—10 mM Tris, pH 8, 1 mM EDTA

3 M potassium acetate, pH 5.5

Method

1. The yeast cultured for this experiment can be at a high density. The yeast can be cultured for as long as 48 hrs at 30°C. The optical density of a 1:10 dilution of the culture in water can be as high as 1.0 at 520 nm.

2. Harvest 5 ml of cells by centrifugation (5 min at 5000 rpm). Resuspend the yeast in 1 ml of cold 50 mM EDTA, pH 8, and transfer to a 1.5 ml microfuge tube. Centrifuge for 1 min, decant, and resuspend again in 50 mM EDTA.

3. Pellet the cells as before and suspend the cells in 1 ml of 50 mM Tris, pH 9.5, 2% 2-mercaptoethanol. Incubate for 10 min at room temperature. Centrifuge and decant.

4. Resuspend the cells in 800 µl of 1.2 M sorbitol, 50 mM Tris-HCl, pH 7.5. The sorbitol acts as an osmotic support and prevents rupture of the cells as the wall is removed. As the yeast cell walls degrade, membranes can easily overextend and rupture.

5. Add 200 µl of Lyticase (500 U/ml in 50 mM Tris-HCl, pH 7.5). Place the cells on a rocker and incubate at 37°C for one hour. Lyticase is a yeast cell wall degrading enzyme isolated from the bacteria *Arthrobacter luteus*.

6. Examine the suspension under a microscope to ensure protoplast formation. As the yeast wall is degraded, the cell membrane can ooze out of the sack. Viewed with phase contrast microscopy, yeast protoplasts are characteristically refractile (or bright) spheres, and yeast cell wall shells appear as gray ghosts (cell walls without membrane and cytosol). Combine 10 µl of 10% SDS with 10 µl of yeast proto-

plasts. Examine the cells under the microscope. The absence of refractile yeast indicates the protoplasts were lysed by the SDS.

7. Pellet the protoplasts by centrifuging at 10,000 rpm for 5 min. Resuspend the cells in 1 ml of 100 mM Tris, pH 7.5, 100 mM EDTA, 150 mM NaCl (lysis buffer). Transfer the cells to a 5 ml polypropylene tube. Add 1 ml of lysis buffer with 2% SDS. Mix and incubate at 30°C for 30 min. Check the cells under a microscope for lysis.

8. Centrifuge the lysate at 5000 rpm for 15 min to pellet cellular debris. Decant the upper phase containing the DNA.

9. Using a pipet, determine the volume of the DNA solution. Add $\frac{1}{10}$th volume (e.g., 100 μl for every ml) of 3 M potassium acetate to the solution. In the presence of Na^+ or K^+, DNA precipitates if mixed with either ethanol or isopropanol. Thus, precipitate the DNA by adding 0.7 volumes of isopropanol. Incubate the DNA at –20°C for 30 min (or overnight if possible). Centrifuge the solution at 7000 rpm for 20 min. The DNA appears as a white pellet. Decant and remove as much moisture as possible, but do not allow the pellet to dry. Once genomic DNA drys, it can be very difficult to resuspend.

10. Resuspend the DNA in 100 μl of TE buffer and freeze.

Plasmid DNA Isolation

Plasmids are small, circular DNA molecules commonly used as vectors, i.e., carriers of foreign DNA. They typically consist of an origin of replication (which allows for their replication), a selectable marker (usually antibiotic resistance), and locations in which to insert the foreign DNA. Isolating and purifying plasmids is a routine event in the molecular biology laboratory. Here you will isolate a plasmid in preparation for cloning.

Materials

E. coli—bacterial host strain harboring a plasmid, grown overnight in LB broth with selective pressure.

MORE...

Lysis Buffer—25% sucrose, 50 mM Tris-HCl, pH 7.6, 10 mM EDTA, 50 µg/ml RNase A, & 4 mg/ml lysozyme. RNase and lysozyme are added just prior to use.

0.2 N NaOH/1% SDS—Prepared the day of use.

3 M potassium acetate, pH 5.5

TE Buffer—10 mM Tris-HCl, pH 8.0, 1 mM EDTA.

Syringe (10 ml)

Method

1. Retrieve the culture of *E. coli* prepared for this experiment. Plasmid DNA isolated from 5 ml of the culture provides sufficient plasmid DNA for this and subsequent experiments. A few milliliters of broth, when cultured overnight, can yield as much as 100 to 500 µg of plasmid.

2. Pellet cells from 5 ml of *E. coli* culture in a polypropylene tube by centrifuging at 7000 rpm for 10 min. Decant the supernatant without disturbing the pellet, and resuspend the cells in 1 ml of TE buffer. Centrifuge for 10 min at 7000 rpm and decant.

3. Add 1 ml of lysis buffer and resuspend the pellet by vortexing. Incubate at 37°C for 5 min in a water bath or incubator. This lysis buffer removes the cell wall from the bacteria. At this point the *E. coli* should appear as a gray to yellow/brown suspension.

4. Add 2 ml of 0.2 N NaOH/1% SDS, mix by swirling, and incubate at 37°C for 10 min. The NaOH and SDS cause the cells to lyse; thus the solution should clarify. As the cells are lysed, the chromosomes are released which results in increased viscosity of the solution.

5. Transfer the tube to an ice bucket and cool for 5 min. Add 1.5 ml 3 M potassium acetate, mix well, and incubate on ice for 5 min. The potassium acetate causes the SDS, protein, and chromosome to precipitate.

6. To remove precipitated cellular debris, filter the solution through a syringe stuffed with a Kimwipe. Simply remove the plunger from the syringe, crumple the Kimwipe, stuff it into the syringe, and pack it gently with the plunger. Avoid contaminating the Kimwipe with

nucleases by handling the Kimwipe with clean gloves. Remove the plunger and pour the cell lysate into the syringe. Using the plunger, gently push the lysate through the Kimwipe, and collect the filtrate in a 5 ml tube. A clear filtrate is needed; thus repeat the filtering if necessary.

7. Transfer 800 μl of filtrate to a microfuge tube and add 0.7 volumes (i.e., 560 μl) of isopropyl alcohol. Cap the tube and mix by inversion. Incubate on ice for 30 min. The amount of DNA in the microfuge tube can exceed 200 μg which is sufficient for the purposes of this laboratory. The remaining filtrate can be stored at –20°C, or, if necessary, precipitated.

8. Pellet the precipitated DNA by centrifuging at 10,000 rpm in a microfuge for 10 min. The DNA appears as a small whitish pellet in the bottom of the microfuge tube. Don't be deceived by the size of the pellet; it may contain a large amount of DNA.

9. Pour the supernatant into a waste bottle, tap out remaining liquid onto a Kimwipe without disturbing the pellet, and dry the pellet by placing the tube in a SpeedVac. If a SpeedVac is not available, the pellet can be air dried by placing the open tube on the bench. Cover the opening of the tube with a Kimwipe.

10. When dry, resuspend the pellet in 100 μl of TE buffer and store in a labelled microfuge tube at 4°C.

Estimation of DNA Concentrations

The determination of the concentration of DNA or RNA in solution is a fundamental task in molecular biology. DNA is usually the limiting reagent in most experiments; therefore, knowledge of its concentration is critical. Restriction endonuclease reactions, ligations, transformations, etc., are all extremely dependent upon the concentration of the DNA. Determination of the concentration of DNA can be estimated either by qualitatively comparing the fluorescence of DNA bands in an agarose gel to a standard or by spectrophotometric means.

Qualitative Estimation

DNA fluoresces in the presence of ethidium bromide, and the intensity of the fluorescence is proportional to its concentration. As such, comparison of relative fluorescence of unknown DNA to known standards can be used as a rough estimation of DNA concentration. This comparison is usually made following the electrophoresis of the standard (e.g., 1 μg of Lambda DNA cleaved with the restriction endonuclease *Hind*III) and the unknown. The mass of 1 μg of a Lambda-*Hind*III standard provides convenient references for comparison (Table 8.2).

An alternative method for estimating DNA concentration relies on the lower limit for visually detecting DNA in a gel. It has been estimated that a band of DNA below 5 ng is not detectable by the human eye. Serially diluting DNA to extinction, followed by electrophoresis, can also be used to measure DNA concentration. This technique is particularly useful if there is more than one species or fragment of DNA in a sample. In this manner, the concentration of individual bands can be estimated.

Quantitative Estimation

DNA, RNA, and protein strongly absorb ultraviolet light in the 260 to 280 nm range. UV spectroscopy can be used as a quantitative technique to measure nucleic acid concentrations and protein contamination.

Table 8.2 : A commonly used electrophoretic standard and the masses of individual fragments

Kb Fragment	μg of DNA/1 μg total (ng)
23.31	480
9.42	194
6.56	135
4.36	90
2.32	48
2.03	42
0.56	12
0.13	3

Nucleic acids absorb strongly at 260 nm and less strongly at 280 nm while proteins do the opposite. The general rules for determining the concentrations of nucleic acids at 260 nm are:

1. 1 OD unit of double-stranded DNA is 50 µg/ml;
2. 1 OD unit of single-stranded DNA is 33 µg/ml; and
3. 1 OD unit of single-stranded RNA is 40 µg/ml.

Proteins absorb strongly at 280 nm where 1 OD unit is 1 mg/ml. When using UV spectroscopy for estimating DNA concentrations, it is very important to remove all protein and RNA from the DNA solution. Good estimations can only be made on clean preparations.

An estimate of the purity of a DNA preparation can be made by measuring the absorbance at both 260 and 280 nm. Pure solutions of nucleic acids will absorb approximately twice as much at 260 nm as at 280 nm. Experimentally, the ratio of 260 nm/280 nm of a pure DNA solution is between 1.8 and 2.0. As protein contamination increases, the ratio decreases. Additionally, the presence of contaminating organic solvents, such as phenol, can affect estimations of concentration and purity.

Materials

UV Spectrophotometer
Quartz or UV compatible cuvettes
TE Buffer
DNA Sample

Method

1. Fill the cuvette with water or TE buffer. Zero the spectrophotometer at 260 nm with this blank.

2. DNA from plasmid and genomic preparations is typically at a concentration exceeding 1 µg/µl. Consequently, DNA is usually diluted before measuring its absorbance. An unfortunate result of this measurement is that the DNA is expended as a result of the dilution. Be sure there is adequate DNA to waste. Start by diluting the DNA

sample 1 µl:999 µl of TE buffer (the dilution can be done directly in the cuvette). Mix the dilution thoroughly.

3. Measure the optical density. Multiply the resulting OD by 50 µg/ml. For a 1:1000 dilution, the mass of DNA is equal to µg/µl.

4. Similarly, the same sample can be measured at 280 nm. A ratio of the $OD_{260\ nm}/OD_{280\ nm}$ is an indicator of DNA purity. A ratio of 1.8 or higher indicates minimal protein contamination.

STUDY QUESTIONS

1. In isolating chromosomal DNA from *E. coli*, the detergent Triton X-100 is used. When isolating plasmid DNA, SDS is used. Why are different detergents used?

2. Investigate the difference between low and high copy number plasmids. What is the mechanism for this difference?

3. Besides ampicillin, investigate alternative selectable markers for bacteria. What markers are available for yeast, fungi, and cell culture?

4. Investigate the differences between plasmids and chromosomes.

FURTHER READINGS

Polisky B (1986): Replication control of the ColE1–type plasmids. In: *Maximizing Gene Expression*, Reznikoff and Gold, eds. Boston: Butterworth

Sambrook J, Fritsch EF, Maniatis T (1989): *Molecular Cloning: A Laboratory Manual*. Plainview, NY: Cold Spring Harbor Laboratory Press

9

Constructing a
Gene Bank

9.1   OVERVIEW

Cloning involves isolating, fragmenting, and combining genomic DNA with a vector, followed by introducing the recombinant molecule into a host where it is replicated. The genomic DNA used in cloning is predominantly chromosomal which is large and fragile (i.e., large polymers break). By necessity, the DNA must be broken into manageable pieces, a process normally accomplished by DNA cleaving enzymes called restriction endonucleases. After the DNA is fragmented, the pieces are linked to a vector to form recombined or recombinant molecules. This population of different genomic fragments linked to vector molecules is called a gene bank or gene library.

The typical method of preparing DNA for a gene bank requires that both the genomic and vector DNAs be cleaved with a restriction endonuclease. The restriction endonucleases used for cloning are site specific DNases that recognize four base or longer sequences and cleave the DNA in or near that sequence. Genomic DNA is cleaved randomly to yield an assorted collection of fragments, while the vector is cut only once in a predetermined location. The two sets of molecules are then combined and enzymatically coupled (ligation) by DNA ligase.

Thus far, you have developed a scheme for the cloning process and have isolated both the vector and genomic DNAs. The next step is to construct a gene bank which requires: (1) cleaving the vector site-specifically with a restriction endonuclease; (2) cleaving the genome randomly with a restriction endonuclease; and (3) linking genomic pieces with the vector. In the laboratory you will use restriction endonucleases for the cleavage of genomic and vector DNAs, electrophoretically examine the DNAs, and then recombine DNA molecules through ligation.

9.2 BACKGROUND

The vector, a DNA molecule that carries foreign DNA, is usually a modified plasmid, bacteriophage, or virus that replicates in a host cell. Vectors replicate by using the host's replication machinery, enzymes encoded by the vector, or a combination of both. Vectors also contain short sequences of DNA that are recognized and cleaved by restriction endonucleases (i.e., restriction endonuclease [cut] sites). The vector is cleaved in a precise location while the genomic DNA is cleaved with a restriction endonuclease into a random assortment of fragments. The vector and genomic fragments are then combined by the enzyme DNA ligase.

Cleaving DNA with Restriction Endonucleases

Restriction endonucleases used in cloning are site-specific DNases isolated from bacteria. Their native role is to protect bacteria from viral infections. Bacteria are susceptible to infection by viruses (correctly known as bacteriophages or phages) and, therefore, bacteria have evolved defense systems. This defense involves an enzymatic mechanism that discriminates between bacterial DNA and the DNA of an intruder. Ideally, phage DNA that enters a cell is chopped into pieces rendering it noninfective.

Restriction endonucleases are one of two defense enzymes associated with what is known as a restriction modification system. The second enzyme is a DNA methylase (i.e., a methylase links a –CH$_3$ to the nucleotide base) that modifies either adenine or cytosine bases at specific recognition sites of the bacterial host DNA. Although there are different mechanisms for restriction modification systems, one system works by methylating host DNA at a specific DNA sequence. The restriction endonuclease is incapable of cleaving the methylated DNA; however, foreign DNA entering the cell, which is unmethylated, is degraded by the host restriction enzyme.

Unfortunately for the bacteria, some phage DNA can slip by the restriction modification defense to initiate a phage infection. Interestingly, once phage DNA replicates in a host, the phage DNA is modified by host methylases which then make the phage resistant to attack by that restriction endonuclease upon subsequent infections of the same strain of bacteria. However, bacterial strains, as well as different species, can possess distinct restriction modification systems.

The useful characteristic of restriction endonucleases is the manner by which they cleave DNA. Restriction endonucleases manufactured and sold for cloning usually recognize and cleave within short sequences of four to eight bases. Although traditionally thought to cleave palindromes, i.e., sequences that read the same on opposite strands, restriction endonucleases can recognize both palindromic and nonpalindromic sequences, and they cleave both inside and outside of the recognition site. The usefulness of palindromic cleaving enzymes is that they generate sticky or cohesive or complementary ends (Table 9.1). Sticky ends are actually single-stranded protrusions at the end of the DNA molecule, which are complimentary to themselves and other molecules with the same overhang.

Many restriction endonucleases cleave DNA to generate blunt ends. Blunt-ended molecules are simply those in which the terminus is double-stranded. The important difference between blunt-end and sticky-end molecules is that blunt-ended molecules do not hybridize to other DNAs.

Most commercially available restriction enzymes are simple to use. For activity, the enzyme usually requires a buffer with Mg^{+2}, a pH of 7 to 8, and NaCl, although some enzymes require KCl. Optimal activity is often at 37°C; however depending upon the source, temperature optima can range from 20° to 70°C.

Cloning involves cleaving both the vector and genomic DNAs with a restriction endonuclease which yields compatible sticky ends, and then using those cohesive ends to recombine the DNAs into a construct or

Table 9.1　Representative Restriction Endonucleases, Recognition Sites, and Products of Cleavage

Enzyme	Source	Recognition Site	Product	
AluI	Arthrobacter luteus	–AGCT– –TCGA–	–AG –TC	CT– GA–
BamHI	Bacillus amyloliquefaciens	–GGATCC– –CCTAGG–	–G –CCTAG	GATCC– G–
DraIII	Deinococcus radiophilus	–CACNNNGTG– –GTGNNNCAC–	–CACNNN –GTG	GTG– NNNCAC–
HindIII	Haemophilus influenzae	–AAGCTT– –TTCGAA–	–A –TTCGA	AGCTT– A–
Sau3A	Staphylococcus aureus	–NGATCN– –NCTAGN–	–N –NCTAG	GATCN– N–

recombinant molecule (Figure 9.1). The vector is cleaved in one location, while the genomic DNA is extensively digested into a pool of fragments. The act of linking the vector and genomic fragments is ligation which is catalyzed by DNA ligase. However, between cutting and pasting, DNA must be analyzed for proper cleavage. The analytical technique used is gel electrophoresis, similar to the technique used with proteins.

Agarose Gel Electrophoresis

Agarose gel electrophoresis is the standard method used to analyze DNA molecules larger than 200 base pairs. DNA in alkaline buffers is negatively charged, due to its phosphate backbone, and thus will migrate to the anode in an electrical field. DNA has an even distribution of phosphates and thus a uniform charge to mass ratio. Factors other than charge affect migration and separation of DNA during gel electrophoresis. These factors include the size of DNA, agarose concentration, DNA concentration, and electric field strength. Linear DNA fragments, i.e., those produced from restriction digests, travel through gel matrices at rates

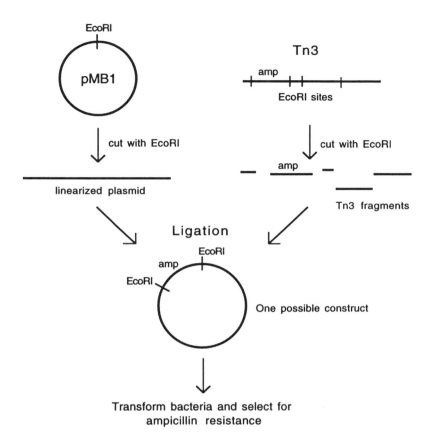

Figure 9.1 Cutting, pasting, and then cloning DNA.

inversely proportional to the $\log_{10}$ of their molecular weight. This relationship can be used to estimate the size of DNA fragments.

Agarose gel electrophoresis is performed primarily with horizontal slab gels. Horizontal gels are made by pouring melted agarose into a gel mold (much like pouring a rectangular single layer cake). A comb hangs into the agarose from the top of the mold to form the wells. After the agarose solidifies, the comb is removed and the gel is placed in the gel

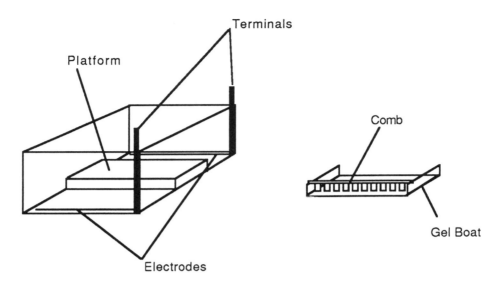

Figure 9.2 Schematic of a horizontal electrophoresis unit.

box. The most common method for electrophoresis of DNA is a submarine method in which buffer completely covers the gel and acts as a heat sink (Figure 9.2).

Vertical electrophoresis systems can be used, especially with very large samples volumes (>20 μl), however, they are difficult to manipulate. Vertical PAGE gels are routinely used for the analysis of extremely small DNA fragments (<500 bp) and oligonucleotides. Analysis of smaller molecules can also be done with horizontal gels using higher concentrations (4%) of highly purified agarose (e.g., Metaphor, FMC BioProducts), although the resolution does not match that of acrylamide gels.

Electrophoresis of DNA is carried out in buffers of neutral or alkaline pH. Under these conditions the phosphate backbone is negatively charged, and thus it migrates toward the anode (+). Generally, buffers with a pKa near the desired pH of the gel and with an ionic strength from 0.05–0.10 M can be used (see recipes below). A pH gradient may develop as buffers with low buffering capacities are depleted during the electrophoresis. If this occurs, the buffer should be replaced or recirculated. EDTA is added to all electrophoresis buffers to chelate divalent cations, which are needed by DNases, and prevents degradation of the sample in the gel.

The following are recipes for two common electrophoresis buffers.

Tris-Borate-EDTA (TBE)

Description: DNA electrophoreses well in TBE buffer at high voltages. Borate buffers have a high buffering capacity and give good resolution, but borate may complex with agarose and interfere with DNA migration. It is difficult to recover DNA from gels run in borate buffers. TBE can be used in DNA gels, sequencing gels, and PAGE.

Formulation: 89 mM Tris, 89 mM boric acid, 2 mM EDTA. A 5X stock is prepared by dissolving 54 g Tris base, 27.5 g boric acid, and 20 ml 0.5 M EDTA (pH 8.0) into water and adjusting the volume to 1 l.

Tris-Acetate-EDTA (TAE)

Description: Acetate buffers are good for recovering DNA from gels and for blotting, but have relatively poor buffering capacity and may require recirculation or replacement of the buffer during longer runs. A good practical aspect of TAE is that the buffer concentrate is 50X.

Formulation: 40 mM Tris-acetate, 2 mM EDTA. To prepare a 50X stock, dissolve 242 g Tris base, 57.1 ml glacial acetic acid, and 100 ml 0.5 M EDTA (pH 8.0) in water and adjust to a volume of 1 l.

When quantitatively analyzing DNA, it is critical that individual bands can be clearly seen, or resolved. Resolution is dependent upon several factors, including the size and shape of the sample well, the size of the DNA fragments, the distribution of the size of those fragments,

Table 9.2   Factors Influencing the Resolution of Bands in Agarose Gel Electrophoresis

Factor	Comments
Well size	Wells with large capacity tend to decrease resolution. Wide wells cause the sample to be diffuse upon entering the gel bed. The best resolution is obtained from combs with thin teeth. Thick teeth are useful for preparative electrophoresis.
Fragment size	The capacity, which is the upper limit of DNA that can be loaded onto a gel without losing resolution, drops sharply as fragment size increases, especially over a few thousand base pairs. Large fragments that are overloaded will result in smearing and trailing bands, which becomes more pronounced with increasing DNA size. In standard agarose gel electrophoresis, fragments greater than 10 Kb are difficult to quantitate and if overloaded, difficult to resolve.
Agarose concentration	As the concentration of agarose increases, the resolution of larger bands drops considerably. Concentrations up to 6% can be used to separate oligonucleotides. Larger molecules, i.e., greater than 10–20 Kb, can only be effectively separated using pulse-field gel electrophoresis.
Distribution of DNA fragment size	Highest capacity is when the DNA fragments have a continuous, even distribution across a wide size range. Fragments that are very similar in size can be extremely difficult to resolve and at times may appear as one band.
Voltage gradient	Higher voltage gradients (i.e., V/cm as determined by the distance between the electrodes) lead to poorer resolution and hence lower gel capacity. At higher voltages, fragments tend to travel as a compact mass.

the concentration of agarose used, and the voltage used for separation (Table 9.2).

Like acrylamide, agarose forms a three dimensional lattice that maintains the shape of the gel but in which buffer and molecules can migrate. Depending on the concentration of agarose, the size range into which DNA can be efficiently separated differs. Furthermore, the linear relationship between DNA mobility and size (log of the molecular weight) is dependent upon the gel concentration. Table 9.3 summarizes the effect of agarose concentration on DNA mobility.

Since agarose acts as a molecular sieve, the conformation of DNA affects its migration and separation during electrophoresis. DNA has three normal conformations, namely, supercoiled (closed covalently circular—ccc), relaxed (nicked or open circular—oc), and linear. Generally supercoiled DNA runs faster than linear DNA which has greater mobility than relaxed DNA (Figure 9.3). Since the rules of DNA mobility are not fixed, it is important to run adequate controls when electrophores-

Table 9.3 The Effect of Agarose Concentration on Gel Electrophoresis*

Agarose (%)	Separation Range (Kb)	Gel Strength
0.3	60–5	weak—do not remove gel from boat
0.6	20–1	moderate—exercise care in handling gel
0.9	7–0.5	moderate—exercise care in handling gel
1.2	6–0.4	strong—gel can be handled with ease
1.5	4–0.2	strong—gel can be handled with ease

*Sambrook et al., 1992

ing DNA. Typically an uncut control is run adjacent to DNA cut with restriction endonucleases.

Field strength (voltage) also determines the means by which DNA migrates. At low voltages, migration of linear DNA is proportional to its mass. As field strength is increased, the mobility of higher molecular

Figure 9.3 Mobility of supercoiled, relaxed, and linear DNAs as measured by agarose gel electrophoresis. (A) Linear fragments from a Lambda-*Hin*dIII digest; (B) Supercoiled (lower) and relaxed (upper) plasmid DNA; (C) Linear DNA.

weight fragments increase differentially, i.e., the fragments in the gel do not move proportionally to the $\log_{10}$ of their molecular weight. Thus, separating DNA becomes less quantitative as voltage is increased. For molecules to migrate proportionally to their molecular weight, gels should be run at no more than 5 V/cm (as determined by the distance between electrodes). At very low voltages, DNA in the bands may actually diffuse while the gel runs, thus reducing resolution.

The fluorescent dye, ethidium bromide, is used for rapid, sensitive staining of DNA in agarose. It contains a planar group that intercalates between the stacked bases of DNA. The ethidium bromide–DNA complex displays approximately tenfold greater fluorescence than free dye. UV radiation absorbed by the DNA at 260 nm and transmitted to the dye, or radiation absorbed at 300 nm and 360 nm by the bound dye itself, is emitted in the red–orange region of the visible spectrum.

Ethidium bromide (0.5 μg/ml) may be incorporated into the gel, and running buffer or the gel may be stained separately after electrophoresis. Incorporation of the dye into the gel decreases the mobility of linear and open circular DNA, and it may have an indeterminate effect on closed circular DNA (depending on the degree of supercoiling). If contamination of equipment with ethidium bromide is undesirable, staining after electrophoresis is the method of choice. **Remember that ethidium bromide is toxic and carcinogenic!**

Stain gels, after electrophoresis is carried out, in 0.5 μg/ml ethidium bromide in distilled water at room temperature for 45 min. This time varies depending on gel thickness. Destaining is usually not necessary; however it reduces background and enhances detection of small amounts of DNA. If required, soak the gel 30–60 min in water with occasional rocking. Prolonged destaining causes loss of dye bound to DNA and diffusion of the bands.

Following staining, the gel is placed on a UV transilluminator (light box). Upon illumination, DNA appears as orange bands. The UV light will damage the DNA; thus if the DNA is to be excised from the gel for subsequent manipulations, limiting the time of exposure is important. Gels are typically photographed with a Polaroid camera fitted with a Yellow 14 filter. This filter excludes the UV light but not the visible light. Fluorescent rulers can be place next to or on top of the gel so that the migration distance of the bands can be accurately measured from the photograph.

Fragmenting Large Genomic DNA for Cloning

Most hosts in cloning are limited from accepting large DNA constructs (i.e., combined vector and foreign DNAs). Consequently, the in vitro recombination of DNAs is not performed indiscriminately. Following the isolation of DNA, it is necessary to cleave the large molecules into smaller manageable pieces of approximately 10 to 15 kilobases. Depending on the source of the DNA, the amount of cleavage will differ. For instance, the *Escherichia coli* chromosome is 4×10^6 bases or 250 times larger than an easily manipulated piece used for cloning. In comparison, the human genome is 2.8×10^9 bases and thus would require more extensive fragmentation. Assuming a fragment size of 10,000 bases, cloning the *E. coli* chromosome would require breaking the molecule into at least 400 pieces while the human genome would require a minimum of 280,000 pieces. As we will see below, the actual number of fragments needed is much greater.

The preparation of DNA for cloning makes use of a restriction endonuclease's specificity in cleaving DNA at defined sequences. These restriction sites occur at random, unknown locations within a DNA molecule, and their frequency depends on the number of nucleotides within the site. A four bases sequence, on average, occurs once every 256 bases, while six bases occur every 4096 bases, respectively. Treating DNA with a four base cutter (i.e., a restriction endonuclease which cleaves a four base recognition sequence) would severely degrade the DNA. However, limiting the amount of enzyme yields a partially degraded genome (not all the sites have been cleaved) of variously sized fragments (Figure 9.4). Even if a restriction site falls within a gene of interest, a partial digest should produce a fragment with the gene intact.

Partial digests are performed by serially diluting an enzyme so that it is limiting and can not cleave all available restriction sites. Thus a serial dilution of a four base cutter would yield fragments on average of 256 bases and up. The degree of digestion of DNA by a restriction endonuclease is measured by agarose gel electrophoresis. By comparing the digested products to known size standards, an estimate of the average size of a digest can be determined (Figure 9.4).

The DNAs loaded onto the agarose gel illustrated in Figure 9.4, with exception of the size markers, do not yield specific bands as do plasmids (Figure 9.3). Since the enzyme is randomly cutting the DNA, the result is

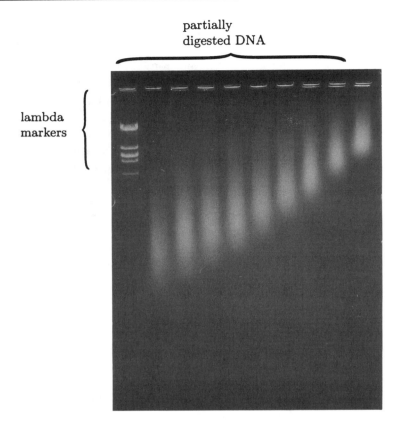

Figure 9.4 Size of partial digest products as determined by agarose gel electrophoresis. The average size of the fragments in each lane can be estimated by comparison to the lambda *Hind*III standard.

a continuous range of fragments, which on a gel appears as a smear. However, note that at the higher concentrations of enzyme, the more intensely fluorescing regions of the smear are associated with smaller DNA molecules. The DNA of desirable size (usually between 10 and 20 kilobases) can be isolated (extracted) from the gel and used for the ligation. Extraction involves removing a block of agarose from the gel and then separating the DNA from the agarose matrix.

The Ligation Reaction

During DNA replication, nicks are created between adjacent 3' hydroxyl on a ribose and the 5' phosphate which is attached to the following ribose (Figure 9.5). This nick is sealed by DNA ligase which, upon expending an ATP, covalently links the two bases. This enzyme is used to covalently recombine DNA in vitro. For cloning, several DNA ligases are commercially available, the most common being T4 DNA ligase isolated from the bacteriophage T4. This enzyme ligates both sticky and blunt-ended DNAs. Also available are a variety of thermostable ligases and a ligase from *E. coli*. These are employed less often than T4 DNA ligase.

Ligation is one of the most difficult techniques to master when working with DNA. The actual enzymatic reaction is simple, but pro-

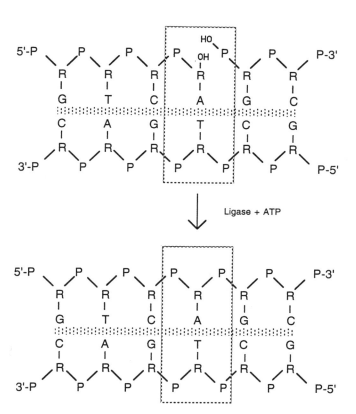

Figure 9.5 The linking of adjacent nucleotides by DNA ligase.

ducing the desired recombinant product is challenging. The problem is not in enzymatically linking DNA, but rather in combining DNA so that a genomic fragment (insert) is linked (inserted) into a vector. When genomic fragments and linearized (i.e., cleaved) vector are mixed and ligated, the products of the reaction are random. The trick is to adjust the concentration of vector and inserts so that a high percentage of ligation products are recombinant molecules. High or low concentrations of DNA can result in very poor ligation results (Table 9.4).

A variety of molecules are produced from a ligation, including monomers (molecules that ligate to themselves), dimers (two different molecules that combine), polymers (three or more molecules), and concatomers (long chains of linked molecules). When linearized vectors and insert concentrations are correct, only 10% to 30% of the ligated products will be recombinant molecules. As such, ligation is a very inefficient process. A technique used to increase the efficiency of forming recombinant products makes use of alkaline phosphatase, an enzyme that can remove the 5' phosphate from the free end of a DNA molecule.

Alkaline phosphatase treated DNAs are incapable of circularizing, i.e., ligating to themselves (Figure 9.6). By removing the terminal 5' phosphates on both ends of a DNA fragment, the ligation of a molecule to itself is impossible. However, a second molecule with terminal 5' phosphates can be ligated to a dephosphorylated vector. The resultant construct would have one nick on both strands, but it would be a recombinant product (Figure 9.6). Such nicked molecules can be introduced into *E. coli* in which the nicks are repaired in vivo and the DNA is replicated. The result of treating a vector with alkaline phosphatase is summarized as follows: (1) phosphatasing a vector prevents its recircularization; (2) inserts can be ligated to phosphatased vectors to create nicked

Table 9.4 Effect of DNA Concentration on Ligation Efficiency

DNA Concentration	Result	Yields
High	Concatomers	low #s of recombinants
Moderate	Some dimers, many monomers	10–30% recombinants
Low	Monomers	low #s of recombinants

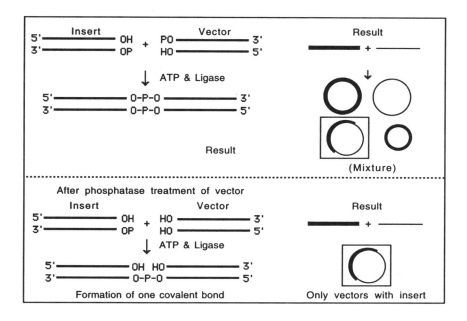

Figure 9.6 Ligation with and without phosphatase treated vector.

recombinant molecules; (3) nicked recombinant molecules can be introduced into *E. coli*; and (4) only *E. coli* that gain recombinant vectors will grow under selective pressure.

Alkaline phosphatase is a relatively easy enzyme to use since it is active in neutral buffers. The most common source is from calf intestine. The major disadvantage to using alkaline phosphatase is that it is difficult to destroy, usually requiring phenol/chloroform extractions, followed by an ethanol precipitation.

9.3   EXPERIMENTAL DESIGN AND PROCEDURES

These laboratory exercises will focus on cleaving DNA with restriction endonucleases, assessing the digestion by agarose gel electrophoresis, and then constructing recombinant DNA molecules by ligation. The ligation products will then be introduced into *E. coli* in Chapter 10.

Cleaving Vector DNA with Restriction Endonucleases

Both donor and vector DNAs must be cut prior to their recombination. Restriction endonucleases are site specific DNases which are used for this cutting. The sticky ends generated by some restriction enzymes allow the vector and donor DNAs to be easily recombined. This experiment will involve cutting vector DNA in preparation for cloning. The vector will be cleaved with the restriction endonuclease *Bam*HI, which will cut the vector in a unique location. The genomic DNA, which will be digested in a subsequent experiment, will yield fragments that have sticky ends to the cleaved vector DNA. The lambda DNA, which is cleaved with *Hin*dIII, will be used as an electrophoresis standard when the vector DNA is examined electrophoretically.

Materials

Restriction enzymes *Bam*HI and *Hin*dIII

10X reaction buffer for *Hin*dIII and *Bam*HI*—(Normally supplied with the restriction enzymes) 100 mM Tris, pH 8, 50 mM $MgCl_2$, 100 mM NaCl, 10 mM 2-mercaptoethanol.

Lambda DNA—electrophoresis marker

Vector DNA—isolated and quantitated previously

3 M potassium acetate, pH 5.5

Phenol:chloroform:isoamyl alcohol under TE buffer—The three solvents are mixed in a ratio of 25 parts phenol to 24 parts chloroform to 1 part isoamyl alcohol. The solution is placed in a brown glass bottle. TE buffer (10 mM Tris, pH 8, 1 mM EDTA) is added to the bottle to saturate the organic phase with water. The organic solvents are the bottom layer.

---

*Different restriction endonucleases require different buffers. Enzymes are usually supplied with their correct buffer.

Phenol will oxidize with time (it turns red) at which point it should be properly discarded. **Take extreme care in handling phenol and chloroform solutions!**

95% ethanol—Ice cold. Keep in the freezer until needed.

TE buffer—10 mM Tris, pH 8, 1 mM EDTA

Method

1. A restriction digest normally involves the cleaving of 1 μg of DNA by 1 U of restriction enzyme (or multiple thereof) per reaction volume. In each reaction recipe listed below, the volumes of water and DNA are omitted. Since the concentration of DNA can vary between individuals, the volume of DNA needed to equal the specified mass will also vary. The volume of water can be adjusted to compensate for individual differences in DNA volume. The overall objective is to mix the correct mass of DNA with buffer and reach the specified volume. As such, set up three small microfuge tubes as follows:

   A. Lambda-*Hind*III digest (electrophoresis marker)

__.__	μl	ddH$_2$O
5.0	μl	10X *Hind*III buffer
.	μl	lambda DNA (5 μg)
49.0	μl	Total

   B. Vector-*Bam*HI digest (digested cloning vector)

__.__	μl	ddH$_2$O
4.0	μl	10X *Bam*HI buffer
.	μl	plasmid DNA (2 μg)
39.0	μl	Total

   C. Vector-undigested (negative control for vector cleavage)

__.__	μl	ddH$_2$O
2.0	μl	10X *Bam*HI buffer
.	μl	plasmid DNA (1 μg)
20.0	μl	Total

2. Restriction endonucleases normally are sold in a concentrate of 10–15 U/µl. An aliquot should be diluted* so that 1 µl of diluted enzyme contains at least sufficient units (U) to cleave the mass of DNA. For instance, *Hind*III should be diluted to 5 U/µl and *Bam*HI diluted to 2 U/µl. Add 1 µl of the appropriate enzymes (at a suitable concentration) to tubes A and B. Flick the tubes to mix the contents.

3. Incubate at 37°C for 60 min.

4. Restriction endonucleases must be inactivated so that they do not cleave any ligations that restore the recognition site. The inactivation of enzymes requires either heat or phenol extraction. Heating the digest to 65°C to 85°C for 30 min will destroy many enzymes. Heat stable enzymes must be inactivated by extracting the sample with phenol/chloroform/isoamyl alcohol. Following extraction, the DNA must be precipitated, washed, dried, and resuspended in TE buffer if a ligation reaction is planned. This is necessary since residual phenol will interfere with the ligation. Inactivation is very important since the digested DNA will be used for ligation procedures.

5. Extract tube B with an equal volume of phenol:chloroform:isoamyl alcohol (25:24:1). **Remember: Chloroform is carcinogenic, and phenol can cause severe burns!** Cap the tube, vortex briefly, and centrifuge to separate the phases. The aqueous layer with the plasmid DNA is on the top. Remove the upper phase and transfer to a new microfuge tube. Properly dispose of the organic phase as indicated by your institution. Add $\frac{1}{10}$ volume of 3 M potassium acetate to the DNA solution and mix. There should be approximately 40 µl of aqueous phase, thus 4 µl of potassium acetate is added. Add two volumes (e.g., 88 µl) of ice cold 95% ethanol and incubate the solution for 30 min or longer. Centrifuge the solution in a microfuge at

---

*In practice, 1 µl of enzyme is usually used for a restriction digest, in which case the enzyme is typically in excess. With some enzymes, such as *Bam*HI and *Eco*RI, extreme overuse of enzyme can cause nonspecific degradation of the DNA. To overcome this, enzymes can be diluted to suitable concentrations. However, the dilution of enzymes from high salt storage buffers to a low ionic strength buffer, as with dilution with water, can cause denaturation. Instructional fact sheets supplied with enzymes typically indicate the proper means for dilution. For *Bam*HI and *Hind*III, dilution in 1× restriction endonuclease buffer is satisfactory.

the maximum rpm for 20 min. The DNA will appear as a tiny white pellet (spot), or may even be undetectable. Decant, dry the pellet, and resuspend in 20 µl of TE buffer. It is advisable to reassess the concentration of this solution with a UV spectrophotometer as described in Chapter 8. Store the DNA solution at 4°C until it is analyzed by agarose gel electrophoresis (subsequent experiment).

Partial Digestion of Genomic DNA

Genomic DNA is typically too large to easily manipulate for cloning. Cutting the genome into small random pieces allows for its cloning and ensures that all the sequences within the genome are represented. By mixing the DNA with diluted restriction endonucleases, the genome is partially digested into a population of molecules which are both manageable and representative of all the sequences within the organism. The enzyme used to prepare the partial digest can be *Nde*II, *Sau*3A, or *Mbo*I, all three of which generate ends that are sticky to the ends generated by *Bam*HI, the enzyme previously used to cleave the vector.

Materials

Yeast genomic DNA containing the *MEL1* gene—isolated previously

*Nde*II restriction endonuclease buffer—100 mM Tris, pH 7.6, 150 mM NaCl, 10 mM MgCl$_2$

Restriction endonuclease *Nde*II (*Sau*3A or *Mbo*I may also be used)

Method

1. Make a cocktail of yeast genomic DNA, 10X buffer, and water as follows: 20 µg DNA, 20 µl buffer, and water up to 200 µl.

2. Label seven microfuge tubes #1–7. Add 40 µl of cocktail to the first microfuge tube and then 20 µl to six subsequent tubes.

3. Add 0.5 µl of *Nde*II (1 U/µl) into tube #1. Mix with a micropipette and with fresh tips, serially transfer 20 µl from tube #1 through tube #6. (This serial transfer, i.e., serial dilution, simply means to succes-

sively transfer 20 µl from tube #1 to tube #2 to tube #3 and so forth. This transfer dilutes the enzyme by 50% after each step. It is important to use a fresh pipette tip for each transfer.) Remove 20 µl from tube #6 and discard. Be certain to leave tube #7 as a negative control.

4. Incubate for 30 min at 37°C and then heat for 10 min at 65°C to inactivate the enzyme. Store the digested DNA at 4°C until needed.

5. Analyze the partial digest by agarose gel electrophoresis (next experiment).

Analysis of Restriction Digests by Agarose Gel Electrophoresis

Throughout the cloning process, DNA must be constantly examined and analyzed. There are several indirect techniques to examine DNA, e.g., gene expression, blotting, UV spectroscopy, but only one direct technique, gel electrophoresis. Both acrylamide and agarose gels can be used to separate DNA; however for large molecules, such as those produced by restriction digests, agarose is the choice. This experiment will analyze the plasmid DNA and genomic DNA cleaved with restriction endonucleases.

Materials

Agarose
1X TAE buffer or 50X TAE concentrate (dilute 10 ml to 500 ml to make 1X TAE)
Restriction digests—Plasmid and partial genomic digests
Horizontal electrophoresis apparatus
Power supply and leads
Loading buffer—0.25% bromophenol blue in 40% sucrose (w/v)
Tape

Method

1. The objective of this exercise is to analyze the restriction digests (plasmid and partial digests of the yeast genomic DNA) you prepared in

a previous experiment by agarose gel electrophoresis. Prepare a 1% agarose solution by dissolving 1 g of agarose into 100 ml of running buffer (e.g., 1X TBE or 1X TAE) using a microwave, burner, or hot plate. Keep the agarose in a 55°C water bath until needed. Tape the ends of a gel boat firmly, place the boat on a level surface, insert the comb, and pour the cooled agarose to a thickness of $\approx \frac{1}{2}$ cm. When the agarose has cooled and solidified (it will be opaque), pour a small volume of running buffer onto the gel surface, specifically around the comb (this prevents the gel from drying out). Gently rock the comb and allow the buffer to seep into the wells (the buffer helps to prevent a vacuum which can rip well partitions upon removal of the comb). Lift out the comb without breaking the well walls. Pour off the overlaying buffer and remove the tape. Place the gel in the electrophoresis box and cover with running buffer. The buffer should just cover the surface of the gel.

2. The number samples to be electrophoresed is limited by the comb used with your horizontal gel. The following arrangement would be ideal for the analysis of the plasmid DNA and genomic partial digests. Numbering the wells from left to right, the gel is loaded as follows:

> Lane 1—Lambda/*Hin*dIII digest
>
> Lane 2—Vector/*Bam*HI digest
>
> Lane 3—Vector uncut
>
> Lanes 4–10—Genomic DNA series partially digested with *Nde*II
>
> Lane 11—Lambda/*Hin*dIII digest

If eleven wells are not available on your gel box, then either run more than one gel or load only alternating samples from the partial digest.

Prior to loading the gel, place an aliquot (10 µl) of lambda-*Hin*dIII digest into the 65°C water bath for 10 min. Lambda phage has a linear genome of 48.6 Kb. At both ends of the molecule are 12 base sticky overhangs called cos sites (for cohesive ends). Cos sites can hybridize intra- or intermolecularly and thus must be separated prior to electrophoresis. Fragments 1 and 4 (Table 9.5) contain the cos region and may hybridize after prolonged storage. Heating the preparation to 65°C for 10 min., followed by quick cooling on ice prior to gel loading, will separate the cohesive ends.

Remove the lambda from the heat and chill on ice. Add ⅕ volume (2 µl) of loading buffer (i.e., ⅕ the volume of the total volume of the sample within the tube) to the lambda, and mix. For each of the other samples to be analyzed, mix 0.5 µl of loading buffer with 2.5 µl of sample. It may be necessary to briefly centrifuge the contents to the bottom of the tube.

Sample loading buffer contains glycerol, sucrose, or ficoll to make the sample more dense than the reservoir buffer. A tracking dye, which runs ahead of most DNA fragments, is also included in the loading buffer. A dye such as bromophenol blue is also a pH indicator and may be useful in detecting samples in which the pH is too low. Color changes occurring during electrophoresis indicate that a pH gradient has formed and the buffer should be recirculated or replaced. A good loading buffer is an aqueous solution of 0.25% bromophenol blue and 40% sucrose (store at 4°C).

3. The DNA/loading buffer solution has a density greater than the running buffer. This makes loading the gel relatively easy. Using a steady hand and a micropipette, hold the tip of a micropipette (with the sample) in the running buffer above the well which is to be loaded. *DO NOT PLACE THE PIPETTE TIP IN THE WELL.* Gently expel the sample from the pipette tip. The sample should sink into the well. Don't be alarmed if some of the sample floats or misses the well. With practice this technique is easily mastered. Load the gel at the cathode (the side of the black lead) and run the DNA toward the anode (run to the red). With a horizontal gel box, leads are often easily switched if the gel is in the wrong orientation. Set the voltage on the power supply to 100 V (or 5 V/cm). Be very careful when performing electrophoresis, as high voltages and significant currents are used. Run the gel until the bromophenol blue dye front is one cm from the end of the gel or as long as time permits.

4. Turn off the power supply and transfer the gel to an ethidium bromide staining bath (0.5 µg/ml). When handling ethidium bromide solutions, **always wear gloves**, and avoid spills. Discard ethidium bromide soiled waste, such as soiled paper towels (or when cleaning

Table 9.5 Size and Molecular Weights of Lambda Fragments
Generated by the *Hind* III Restriction Endonuclease

Fragment #	Size (base pairs)	Mol Wt ($\times 10^6$ daltons)
1	23,130	15.27
2	9416	6.21
3	6577	4.33
4	4361	2.88
5	2322	1.53
6	2027	1.34
7	564	0.37
8	125	0.04

up an inadvertent spill) in a suitable manner.* Stain the gel for at least 10 min and observe the gel using a UV transilluminator. Agarose gels can be documented using a Polaroid camera with a Yellow 14 filter. Specifically designed, fixed focus cameras are convenient and produce the best quality photographs.

5. The size of linear DNA fragments can be estimated by comparing the mobility of fragments of unknown size to fragments of known size. By plotting the $\log_{10}L$ versus M (L = molecular weight or size (Kb) of the DNA and M = mobility, i.e., from well front to dye front) a standard curve is generated and used to estimate the size of unknown fragments. Mobility is simply the distance traveled by the DNA divided by the distance traveled by the bromophenol blue dye.

Lambda-*Hind*III digested DNA fragments are used as a typical standard for estimating DNA fragment sizes. This standard is usually loaded on both sides of the unknown. Table 9.5 shows the size and molecular weight of fragments of lambda-*Hind*III.

---

*Ethidium bromide should be discarded as dictated by your institution. However, items that are contaminated with ethidium bromide, e.g., a lab bench, are often treated with diluted bleach. The bleach reduces the mutagenicity of the ethidium bromide, but does not eliminate it. Take the time to identify the proper guidelines for disposal as determined by your institution.

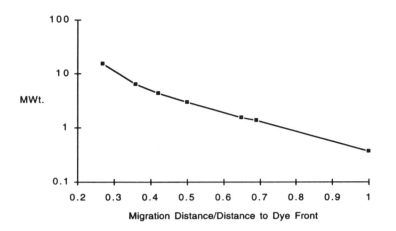

Figure 9.7 Plot of fragment size vs. migration distance for Lambda-*Hind*III digest at 5 V/cm.

Since each band is present in an equimolar amount, fragment 8 is visible only when the gel is overloaded. The linear relationship of mobility versus fragment size is illustrated in Figure 9.7.

Ligation Reaction

Ligation is a complex step in the cloning process. It is difficult to correctly proportion plasmid and genomic DNAs so that a major product is the recombinant molecule. This guide can be used to assist in preparing ligation reactions. Several components are essential in ligations, namely ligation buffer, ATP, linear DNA, and ligase. Buffers are often supplied with DNA ligase and defined by the manufacturers. ATP is usually added to the ligation reaction just prior to enzyme addition. ATP can be made in a stock solution (10 mM in Tris-HCl, pH 7.5) and diluted for use. The DNAs and ligase need to be used in correct proportion to avoid insufficient ligation or the formation of tandem repeats of vector and/or inserts.

For ligations of sticky-ended DNAs, the following generalized conditions should be met:

(1) total DNA concentration of 20 ng/µl (400 ng/20 µl);
(2) 0.5 U ligase/20 µl reaction;

(3) a 2:1 molecular ratio of insert to vector DNAs; and

(4) an overnight incubation at a temperature of 4°C to 16°C.

If phosphatased vector is used, then the ratio of insert to vector should be changed to 1:2.

Blunt ended ligations require:

(1) a DNA concentration of 50 ng/μl;

(2) 5 U ligase/10 μl solution;

(3) a 3:1 ratio of insert to vector; and

(4) an incubation temperature of 20°C for 2 hr.

Materials

Yeast genomic partial digest

Vector DNA—cut with *Bam*HI

10X ligation buffer—A typical 10X ligation buffer is 600 mM Tris, pH 7.5, 50 mM MgCl$_2$, 10 mM dithioerythitol, 10 mM ATP

T4 DNA ligase—1 U/μl.

Method

1. You will be ligating the partially digested yeast genomic DNA into your vector. These molecules have cohesive ends (i.e., sticky ends); therefore set up your reaction as described above. Following ligation, we will transform *E. coli* with the newly constructed gene bank.

2. In three sterile microfuge tubes, mix the following ligation reactions. The volume of plasmid DNA, insert DNA, and water should be based on the concentrations of the DNA solutions and the masses needed as described by following figure (Figure 9.8). The values in this table are calculated based on the size of both the plasmid and insert. Simply cross match the size of your plasmid to the average size of your insert to determine the amounts of DNA needed. You may decide which insert size to use (based on your partial digest gel); however 10 Kb fragments are generally good. A more detailed account of fragment

size will be covered in Chapter 11. The calculations used to construct Figure 9.8 are beyond the scope of this manual.

    A. Reaction with ligase:

_._	μl	ddH$_2$O
_._	μl	yeast DNA digest
2.0	μl	10X ligation buffer
	μl	vector DNA digest
19.0	μl	Total

    B. Reaction without ligase:

_._	μl	ddH$_2$0
_._	μl	yeast DNA digest
2.0	μl	10X ligation buffer
	μl	vector DNA digest
20.0	μl	Total

    C. No DNA:

17.0	μl	ddH$_2$0
2.0	μl	10X ligation buffer
19.0	μl	Total

3. To tubes A and C, add 1 μl (1U) T4 DNA ligase. Keep the enzyme on ice or in the freezer until needed.

4. Cap the tubes, mix, and incubate at 4°C to 15°C overnight.

5. Store these tubes at 4°C until the transformation experiment.

	Insert (bases)			
Plasmid (bases)	5000	7500	10,000	12,500
3000	292/86*	269/53	248/37	231/27
4000	261/103	248/65	233/46	219/35
5000	236/116	230/75	219/54	208/41

*ng insert/ng plasmid for a 20 μl ligation reaction

Figure 9.8 Masses of plasmid and insert DNAs used for a complementary-ended ligation.

STUDY QUESTIONS

1. Based on Table 9.1, what is the major difference between ligating a *Bam*HI cleaved vector to *Bam*HI cleaved insert and a *Bam*HI cleaved vector to *Sau*3A cleaved insert?

2. Following the construction of a gene bank, you introduced the recombinant molecules into *E. coli*, and the bacteria failed to grow. Assuming that the transformation was not faulty, what are the possible sources of error for the construction of the gene bank?

3. Propose a simple experiment to test whether a sample of DNA ligase still contains active enzyme.

FURTHER READINGS

Jackson D, Symons R, Berg P (1972): Biochemical method for inserting new genetic information into DNA of Simian Virus 40: Circular SV40 DNA molecules containing lambda phage genes and the galactose operon of *Escherichia coli*. *Proc Nat Acad Sci USA* 69:2904–2909

Perbal B (1984): *A Practical Guide to Molecular Cloning*. New York: John Wiley & Sons

REFERENCES

Sambrook J, Fritsch EF, Maniatis T (1989): *Molecular Cloning: A Laboratory Manual*. Plainview, NY: John Wiley & Sons

10

Introducing Recombinant Molecules into *Escherichia coli*

10.1 OVERVIEW

The construction of a gene bank is only part of the cloning process. To clone DNA, the recombinant molecules must be introduced into an organism in which they are replicated, and as the cell grows and divides, also propagated. In genomic cloning, the recombinant DNA is normally introduced into the bacterium *Escherichia coli*.

Many organisms can act as hosts for recombinant DNA, but *E. coli* is the most common. Chemically treated *E. coli* take up extracellular DNA in a process called transformation. Once the DNA is within the cell, the vector provides necessary information for the recombinant molecule to replicate and survive. As noted previously, a vector has an origin of replication (ori) and selectable markers, such as antibiotic resistance genes, both of which ensure the plasmid's replication and survival.

In Chapter 9 you created a gene bank by linking fragmented yeast genomic DNA into linearized vectors. By adjusting the concentrations of the DNA within the ligation reaction, a percentage of the ligated products are recombinant molecules containing both plasmid and yeast DNA.

However, this pool of ligated DNA molecules contains recircularized vectors in addition to other products.

In this experiment, you will introduce the ligated DNA into *E. coli.* This process first requires the preparation of competent *E. coli* to take up the DNA. Second, it requires performing the transformation, and, finally, selecting for transformed bacteria. Furthermore, a strategy will be employed that differentiates between cells harboring recircularized and recombinant plasmids.

10.2   BACKGROUND

The process of introducing DNA into a bacterium, such as *E. coli*, is called transformation, and it dates back to the 1920s when genetic material was transferred from dead *Streptococcus* to living cells by Fred Griffith. Although the concept and methodology for transformation existed for over 50 years, it wasn't applied to recombinant molecules until after scientists were able to cleave DNA with restriction endonucleases and then recombine the pieces with DNA ligase. The first in vitro synthesis of recombinant DNA was performed in 1972, while the introduction of recombinant DNA into *E. coli* was first reported in 1973 (Cohen et al., 1973). Since then, techniques for the introduction of DNA into many different bacteria, yeast, fungi, protozoa, plant cell culture, animal cell culture, plants, and animals have been developed.

Transformation Methodologies

Since the initial experiments in cloning, many different methods have been used to transform *E. coli*. The ability of *E. coli* to assimilate extracellular DNA is dependent upon the strain and treatment of *E. coli*. By suspending actively growing *E. coli* in cold $CaCl_2$ the cell becomes permeable or competent. Competent cells are those that are capable of assimilating exogenous DNA. Added DNA is believed to adhere to the cell wall. Heat shocking the cells causes swelling of the cell, and the DNA is dragged in with an influx of water (Figure 10.1).

Only a small percentage of cells assimilate DNA during transformation, i.e., as little as 0.1% of the cells. This low efficiency is overshadowed

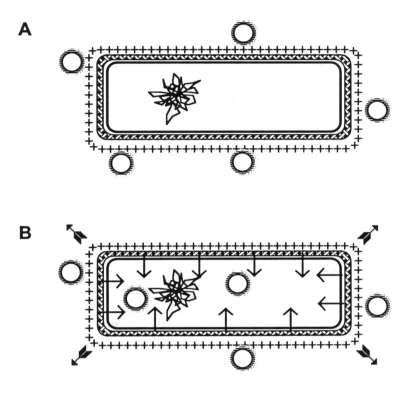

Figure 10.1 (A) The transformation process. Calcium is believed to modify the cell envelope making it permeable to DNA. The cells are mixed with DNA and incubated on ice. (B) The transformation process. Heat shocking the *E. coli* (i.e., 90 sec at 42°C) causes the cells to swell, during which DNA passes into the cell.

by the yield because extremely high cell densities compensate for low transformation efficiencies. For instance, 10^9 cells can yield 10^6 transformants from as little as 1 µg of DNA.

The basic calcium chloride protocol for transformation has been modified and fined tuned by many researchers. Although many strains of *E.coli* are easily transformed, not all strains become competent by treating them with calcium ions.

Many cells, specifically, fungi and plants, require removal of the cell wall prior to transformation. The transformation is based on removing

Table 10.1   Enzymes Used for Cell Wall Removal

Enzyme	Source	Works Against	Specificity
Lysozyme	chicken eggs	bacteria	peptidoglycan
Glusulase	snail gut	yeast	mannans
Lyticase	*Arthrobacter*	yeast	mannans
Novozyme	*Trichoderma*	mold	chitin
Cellulase	*Trichoderma*	plant cells	cellulose

the cell wall in an isotonic solution, followed by the addition of DNA, and finally, by fusion of cells by the addition of polyethylene glycol (PEG). The PEG causes the membranes of cells to fuse, creating giant cells, and in the process, DNA between the cells is assimilated. Cell wall removal is usually accomplished enzymatically with carbohydrases (Table 10.1).

For animal cells, DNA uptake is termed transfection as opposed to transformation. With cell culture, transformation indicates the cells have acquired an immortal or continuous growth characteristic. Several different techniques are available for transfection. Precipitation methods involve mixing the DNA with calcium phosphate or DEAE-dextran which precipitates the DNA. The cells then engulf the precipitate (i.e., endocytosis). An alternative method employs lipid-like molecules that encapsulate the DNA and are then fused with the cell membrane. Generally these methods are much less efficient than *E. coli* transformations (Table 10.2).

Aside from chemical treatment, other physical methods have been developed for transformation, such as electroporation, microinjection, and microprojectiles (biolistic). Electroporation is gaining rapid acceptance because it is easy and efficient. In this technique, cells are mixed with DNA in a cuvette-like container containing two electrodes (Figure 10.2). The cells are then electrocuted (killing approximately 50%) and in the process, DNA is pulled or electrophoresed into the cell. Where 1 µg of DNA could yield 10^6 transformants, electroporation could easily yield 10^8 or higher. Although used with *E. coli*, electroporation is used on other cells as well, such as tissue culture, plant protoplasts, and fungi.

Microinjection is a precise but inefficient technique for introducing nucleic acids into large cells. Using a microscope and a microsyringe with forged glass needles, large cells such as frog eggs and stem cells can be individually injected with DNA (or RNA). Manual injections are exact yet slow and tedious. Microinjection is applicable to the development of transgenic animals (i.e., animals in which DNA has been permanently introduced into the *species*) or for analysis of protein produced from mRNA. For transgenic animals, DNA is injected into embryonic stem cells which are then inserted into young blastocysts in vitro. The blastocysts are then implanted into surrogate mothers (Figure 10.3). Microinjection is also used to inject mRNA into frog eggs (oocytes) in which the mRNA is translated. Integral membrane proteins are often studied using oocyte translation techniques since proteins can be synthesized in a semiisolated, defined system.

A novel method used for cell transformation makes use of microprojectiles, also known as biolistics. The method is analogous to coating microscopic shotgun pellets with DNA and then blasting the cells. As pellets pass through the cell, DNA is left in their wake. Microprojectiles are useful for the transformation of difficult to transform organisms or for the introduction of DNA into organelles, such as the mitochondria and chloroplasts.

An important note about transformation and transfections is that not all species, cell types, or strains transform equally well. Strain to strain

Table 10.2 Transfection Techniques for Cell Culture

Technique	Comment
Calcium Phosphate	When combined, calcium phosphate and DNA precipitate. When layered over adherent cells, the precipitate is assimilated. This technique requires large amounts of DNA, but is generally efficient.
DEAE-Dextran	The dextran polymer is believed to complex with DNA and promote its endocytosis. DEAE-dextran is toxic to many cell lines.
Cationic Lipid-Mediated	Lipid analogs form liposomes (i.e., small membrane vesicles) that encapsulate the DNA. The liposomes fuse with cell membranes at which time the DNA is transferred into the cell.
Electroporation	The efficient technique relies on the ability of an electric potential to create small pores in plasma membranes. These pores are temporary and are the location of DNA entry into the cell.
Protoplast Fusion	Protoplasts prepared from either bacterial cells or minicells (i.e., small cells that break off from *E. coli*) containing plasmid DNA. These protoplasts can be fused to tissue culture with polyethylene glycol.

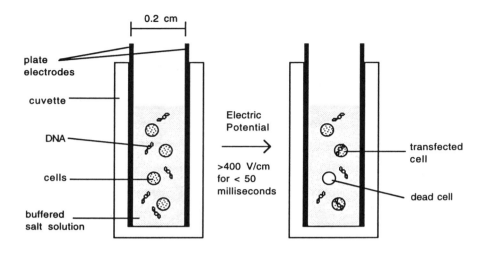

Figure 10.2   Mechanism for electroporation.

or species to species variation can result in tremendously different transformation efficiencies. For instance, laboratory strains of *S. cerevisiae* are
readily transformable; however, industrial strains can be virtually impossible to transform. As such, it is important to be aware that transferring
genes from Donor A to Host B may often be prevented by the inability to
get the DNA into the new host.

Selection of Transformed Organisms

Introducing recombinant vectors into *E. coli* allows for the propagation
of those molecules, but it does not identify the molecules of interest.
Since a typical gene bank may comprise hundreds of thousands of clones,
it is necessary to devise selection strategies that will allow the identification of the desired clone. Selection strategies have two tiers; first is a
general selection for recombinant vectors, and the second is selection
focusing on the specific clone. This section will discuss both but will
emphasize on selecting recombinant vectors while Chapter 11 will concentrate on clone selection.

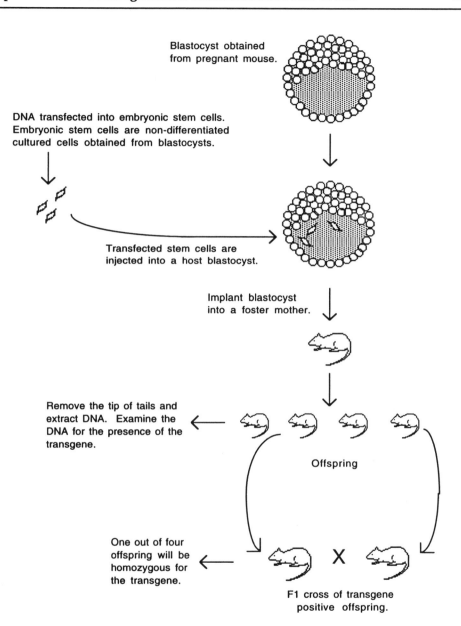

Figure 10.3 Steps in creating a transgenic animal.

Since not all cells assimilate exogenous DNA, it is necessary to screen for those that do. The initial screening of cells begins with differentiating between those that are transformed and those not transformed. With *E. coli* this is relatively easy since vectors normally contain resistance genes that render transformed cells resistant to antibiotics (e.g., ampicillin, tetracycline, kanamycin, neomycin, streptomycin). Transformed *E. coli* simply require culturing on media containing antibiotics. Nontransformed cells die or become stagnant while those with vectors survive and grow.

Not all organisms are sensitive to antibiotics, so some require different selective pressures. For instance, yeasts are readily mutated to be auxotrophic (i.e., incapable of synthesizing an essential metabolite) which can be used as a selection. Plasmids that harbor a gene complementing the auxotrophic mutation can provide the essential factor for the yeast. Growing the yeast on a minimal media, i.e., one missing the essential nutrient, makes the plasmid a necessity for survival. In yeast, plasmids that complement the auxotrophic mutants of *ade*1, *leu*2, *trp*1, and *ura*3 are commonplace.

In cell culture, auxotrophic mutants are rare; however, one commonly used marker is hgprt (i.e., hypoxanthine guanine phosphoribosyl transferase) whose absence prevents purine synthesis. The *E. coli* xprt (xanthine phosphoribosyl transferase) has been modified to express in animal cell culture and thus can be used as a plasmid based selectable marker to offset the genomic mutation. Alternatively, one antibiotic that can be used as a selectable marker with many cells (from bacteria to cell culture) is Antibiotic G418, also known as Geneticin. This aminoglycoside inhibits protein synthesis, but it is rendered harmless by the gene products of the bacterial Kan^R (kanamycin resistance) and Neo^R (neomycin resistance) genes. When these genes are placed downstream of a host promoter, for instance the SV40 early promoter (i.e., a strong promoter from Simian Virus 40), they can be expressed in foreign hosts and yield active enzymes to inactivate Antibiotic G418. This same strategy is used with the antibiotic hygromycin.

Although cells with vectors thrive, in no way does this selection differentiate between those cells with circularized vectors and those with recombinant vectors. Ligation of vector and inserts yields many products, such as: (1) closed vectors (circularized); (2) circularized inserts; (3) vectors linked to vectors; (4) inserts linked to inserts; and, most importantly, (5) inserts linked to vectors. As indicated, ligation is a random linking of DNA fragments. When the ligation products are introduced into *E. coli*, only cells with the vector sequences can survive the selective pressure. That vector, however, may or may not contain an insert. Other nonvector DNA molecules may enter *E. coli* but fail to provide antibiotic resistance, and hence the bacterium fails to survive.

For cells which grow under selective pressure, several mechanisms are available to aid in the identification of cells that acquire recombinant vectors. The early cloning experiments used complementation to select for recombinant vectors. For instance, Ratzkin and Carbon (1977) inserted yeast genomic fragments into a vector, transformed a *leu*B *E. coli* mutant, and plated the bacteria on leucine deficient minimal media. Surprisingly, *E. coli* that picked up the yeast *LEU2* gene were able to grow. This example of complementation (i.e., using *LEU2* from yeast to offset the *leu*B mutation in *E. coli*) is both the direct selection of a specific gene and a recombinant plasmid. This experiment is also the exception rather than the rule since most genes will not express if simply transplanted from one organism to another. Usually an intermediate step of identifying recombinant vectors is required.

Identification of Recombinant Vectors

The standard method for identifying a recombinant vector is by gene inactivation. This involves ligating DNA into a gene which results in the destruction of that gene and loss of a phenotype (i.e., the characteristic imparted by the gene). pBR322 is a vector that is a good model for gene inactivation and has a prominent history. It was constructed by Bolivar et al (1977). Aside from the origin of replication, pBR322 also contains ampicillin and tetracycline resistance genes, each of which contain several unique restriction endonuclease sites (Figure 10.4). Ligation of DNA into either gene inactivates that gene and the corresponding antibiotic resistance. Upon introducing the ligation mixture into *E. coli*, selecting for transformants is accomplished by plating the cells on media containing the antibiotic corresponding to the resistance gene, which is still intact. Only cells that contain vectors will grow. The identification of recombinant vectors is then attained by replica plating the transformants onto media containing the antibiotic corresponding to the inactivated resistance gene. Replica plating is a technique in which an impression of cells from one petri dish is transferred to another dish. Cells that grow on the

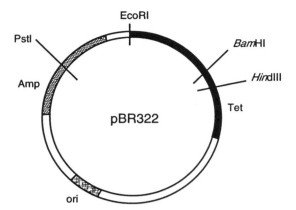

Figure 10.4   The plasmid pBR322.

replica plate possess both active resistance genes and thus are not recombinant vectors (Figure 10.5). This two step strategy is laborious compared to alternative approaches (see below).

A second approach also uses gene inactivation, but only requires one step. The *E. coli* gene *lacZ* codes for the periplasmic enzyme β-galactosidase, one of three enzymes encoded by the *lac* operon. Though

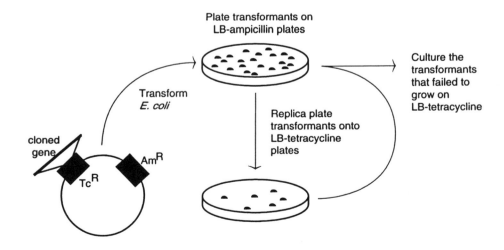

Fig. 10.5   A two step selection for recombinant transformants.

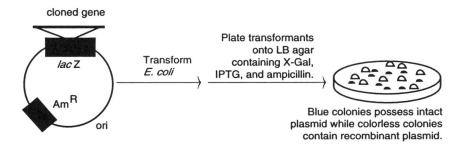

Figure 10.6   Indirect selection for recombinant transformants.

E. coli normally uses β-galactosidase to cleave lactose, the enzyme will hydrolyze any β-galactoside, including the chromogen (color yielding substrate) 5-bromo-4-chloro-3-indoyl-β-D-galactoside (X-Gal), a colorless derivative of the blue dye indole. *E. coli* have been constructed that lack β-galactosidase activity (i.e., *lac*Z mutants). Plasmids with wild type *lac*Z that are introduced into *E. coli* confer the β-galactosidase phenotype. When cells synthesizing β-galactosidase are grown on media containing X-Gal and the *lac*Z inducer IPTG (isopropyl-β-D-thiogalacto-pyranoside), the enzyme releases the indole derivative from the galactose. This 5-bromo-4-chloro-indole is both blue and insoluble, so cells that cleave X-Gal deposit a blue precipitate around themselves. By cloning into the *lac*Z gene, cells that acquire recombinant plasmids are colorless while those with intact vectors turn blue. Specially designed vectors include a multiple cloning site region that contains many unique restriction sites within *lac*Z. This indirect differentiation of cells is an efficient and widely used technique in cloning (Figure 10.6).

Yet another gene inactivation strategy for selecting recombinants makes use of the streptomycin sensitivity gene. *E. coli* resistance to streptomycin is based on the lack of the *str*A gene. Wild type *str*A confers streptomycin sensitivity while the absence, mutation, or inactivation *str*A gene confers resistance. The selection scheme (Figure 10.7) uses a streptomycin resistant (*str*A⁻) *E. coli* and a plasmid with both *str*A and Am^R (ampicillin resistance). *E. coli* harboring the plasmid are sensitive to streptomycin, due to *Str*A, but resistant to ampicillin. The *E. coli* is resis-

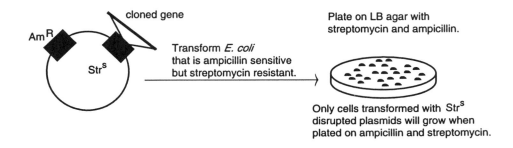

Figure 10.7   Direct selection for recombinant transformants.

tant to streptomycin but sensitive to ampicillin. However, by inserting DNA into the plasmid's *str*A gene, and then introducing the recombinant vector into the *E. coli*, the bacterium now becomes resistant to ampicillin (due to the vector) and resistant to streptomycin (due to the lack of an active *str*A). As a result, transformed *E. coli* that grow when plated on media containing both ampicillin and streptomycin must contain recombinant plasmids.

10.3 EXPERIMENTAL DESIGN AND PROCEDURES

Introducing DNA into *E. coli* is a two step process. The first step is to make the cells competent for the uptake of DNA, while the second step is the actual transformation. In this laboratory, you will prepare competent cells and then transform those cells with the ligation reactions you prepared in Chapter 9.3.

Preparation of Competent Cells

Several methods are available for the treatment of cells to make them competent. The simplest of these, but still reliable, involves washing the cells with a cold calcium chloride solution. Once the cells are treated, they remain competent for 36 hours or longer.

Materials

> *E. coli* culture (such as DH5α or TOP10F' grown overnight)*
> LB broth—50 ml in a 250 ml flask; autoclaved
> Calcium chloride solution (50 mM)—refrigerated; sterile

Method

1. Preparing cells for transformation requires at least three days. On day one, the *E. coli* need to be activated or streaked onto LB agar plates and then cultured at 37°C. On day two, the *E. coli* from an isolated colony on the streak plate is inoculated into LB broth and cultured overnight at 37°C with agitation. Then on day three, inoculate fresh LB broth (50 ml) with 100 μl of the overnight culture of *E. coli*. Incubate on a shaker at 37°C until cloudy (e.g., OD = 0.6). This will take approximately 2 to 3 hours.

2. In a 5 ml tube, centrifuge 4.5 ml of culture, decant the supernatant, and resuspend the pellets in 2 ml of cold 50 mM $CaCl_2$. Immediately centrifuge the cells, decant, and suspend the pellet in 400 μl of $CaCl_2$. Incubate the culture on ice for 20–60 min. The $CaCl_2$ will make the *E. coli* susceptible to pick up DNA. Dispense 100 μl into four microfuge tubes and place on ice.

3. At this point the cells are competent for transformation and can be refrigerated for up to 36 hours without significant loss of transformation efficiency.

Transformation of *E. coli*

The transformation of competent *E. coli* is routine and easy. It involves mixing DNA and cells, incubating the cells on ice, and then heat shocking the cells for 90 sec at 42°C. LB broth is then added to the transformed

---

*The *E. coli* used in this experiment can be the same used in the plasmid isolation, however, it *must lack a plasmid*. The strain DH5α is readily available while TOP10F' can be obtained through Invitrogen.

cells to allow the bacteria to recuperate. The bacteria are then spread plated onto selective agar and incubated. Cells that assimilated recircularized or recombinant plasmids appear within 24 hours.

Materials

Competent *E. coli*—prepared earlier

Ligation solutions—prepared earlier

Positive control—plasmid solution or previously prepared gene bank.

LB broth

Selective agar plates—At least four LB agar with ampicillin, IPTG, and X-Gal. These plates contain 2% agar, 1% tryptone, 1% NaCl, 0.5% yeast extract, and 4 mg/100 ml X-Gal (5-bromo-4-chloro-3-indoyl-β-D-galactoside). Autoclave the agar, cool to 55°C and then add filter sterilized IPTG (isopropyl-β-D-thiogalactopyranoside) and ampicillin. Use 1 μl/ml of 1 M IPTG and 2 μl/ml of ampicillin (25 mg/ml).

LB agar plate

Method

1. Retrieve four microfuge tubes with 100 μl of competent cells. Keep the tubes on ice at all times. Label the tubes to correspond with the ligations you prepared during the earlier exercise (p. 227) and label one as a positive control.

2. Add 10 μl of the ligation solutions and the positive control to its corresponding tubes, mix gently, and incubate on ice for 30 min.

3. Heat shock the cells at 42°C for 90 seconds in a water bath or heat block, and then return the microfuge tubes to the ice.

4. Add 0.9 ml of LB broth to each tube. Incubate at 37°C for 30 min.

5. Transformants will be selected on LB/Ampicillin/X-Gal/IPTG agar and LB agar (no selection). The four DNA solutions (ligation reactions and positive control) are to be plated onto LB/Am/X-Gal/IPTG,

while ligation Reaction A is further plated on LB. (Plating an aliquot from Reaction A onto LB simply demonstrates the cells are viable.) Spread plate with 100 μl aliquots of the transformants onto the agar media. Spread plating involves smearing an aliquot of the cells onto the surface of a agar plate with a hockey stick, i.e., a bent glass rod. Pipette 100 μl of cells onto the surface of the plate. Dip the end of the hockey stick into alcohol and flame. Be careful to avoid the flaming alcohol. With the sterilized hockey stick, spread the aliquot of cells completely over the surface of the plate. Cover the plate and flame the hockey stick to destroy any residual microbes. The remaining transformants can be refrigerated for several days if it is necessary to plate additional transformants.

6. Incubate the plates in a 37°C incubator overnight (be sure to invert plates so that condensation will not smear your colonies).

7. The following day, check your plates for growth. Two types of colonies should be observed. The majority of colonies should be blue which represents recircularized vector. A smaller percentage should be colorless or white. These colonies represent cells with recombinant plasmids. It is important that these plates *do not* remain in the incubator for an extended period of time. Over time, β-lactamase will diffuse from the colonies and destroy the ampicillin in the media. This will allow the growth of nontransformed cells as the ampicillin concentration decreases. If necessary, refrigerate the plates until time permits examination and screening.

STUDY QUESTIONS

1. Investigate the characteristics of vectors that are used with organisms other than *E. coli*. For instance, what are the characteristics of yeast vectors? What about animal cell culture vectors?

2. How is it important to match the characteristics of a particular *E. coli* (Table 1.3) to a specific plasmid?

FURTHER READINGS

Glover DM (1985): *DNA Cloning—A Practical Approach*; Vol. 1. Washington, DC: IRL Press

Sambrook J, Fritsch EF, Maniatis T (1989): *Molecular Cloning: A Laboratory Manual*. Plainview, NY: Cold Spring Harbor Laboratory Press

REFERENCES

Cohen S, Chang A, Boyer H, Helling R (1973): Construction of biologically functional bacterial plasmids *in vitro*. *Proc Nat Acad Sci USA* 70:3240–3244

Ratzkin B, Carbon J (1977): Functional expression of cloned yeast DNA in *Escherichia coli*. *Proc Nat Acad Sci USA* 74:487–491

Bolivar F, Rodrigues RL, Greene PJ, Betlach MC, Heyneker HL, Boyer HW, Crosa JH, Falkow S. (1977): Construction and characterization of new cloning vehicles. II. A multipurpose cloning system. *Gene* 2:95–113

Chapter

11

Screening for Clones

11.1 OVERVIEW

Once *E. coli* is transformed with a ligation mixture, the task of identifying the desired clone follows. Depending on the DNA source, tens to hundreds of thousands of clones may be screened before the target is identified. As we explored in Chapter 7, the strategy used in the cloning process will determine the amount of work required to identify that clone.

Cloning strategies usually include the use of probes to search *E. coli*. These probes may be oligonucleotides derived from the amino acid sequence of a protein, a closely related gene cloned from a different species, or cDNA synthesized from the same species. Regardless of the origin, probes have the common feature of being homologous to, and thus being able to hybridize with, the targeted gene. Probes labelled with a radioactive tag or with an antigen can be used to rapidly screen thousands of clones to identify the desired gene.

The isolation, fragmentation, and ligation of DNA, and the subsequent transformation of *E. coli* were all steps necessary to generate a large population of random clones. The objective of this chapter is to provide you with the techniques needed to screen these random clones in order to find the *MEL1* gene. The laboratory section of this chapter will involve synthesizing a labelled probe, testing the probe, and applying that probe to screen *E. coli* colonies harboring recombinant DNA.

11.2 BACKGROUND

The insertion of genomic DNA/vector constructs into *E. coli* yields an unordered population of transformants which you hope will represent all the DNA sequences of that organism. Complete representation may require hundreds of thousands to millions of clones. The overwhelming number of clones in a library makes the subsequent step of identifying a particular clone a very laborious process indeed. Various approaches can be used for screening, but most involve probes, i.e., labelled DNA with homology to the targeted gene.

Depending on the origin of genomic DNA, the number of clones that must be screened will differ. As we will see below, screening a bacterial gene library could involve examining as few as 2000 colonies while a human genomic DNA library would require the analysis of as many as 7×10^6 clones. Hence, the size of a genome greatly affects the screening process, with smaller genomes requiring less work. Genomes range in size from several thousand bases for a simple phage to several billion bases for higher eukaryotes (Table 11.1).

Parameters of Screening

When a genomic library is constructed and introduced into *E. coli*, the initial result is hundreds to thousands of petri dishes each covered with as many opaque colonies. It is easy to become overwhelmed by the notion that all but a couple of colonies are not immediately valuable. From this large pool, the number of clones screened is primarily dependent upon the size of the genome and the size of the fragments produced in the partial digest (Chapter 9). Researchers have no control over the genome size, but they do control the fragment size. As noted earlier, random genomic fragments are generated by partially digesting DNA with a restriction endonuclease (Chapter 9). Fragments of a desired size can be

Table 11.1 Representative Genomes and Screening Requirements for 10Kb Fragments with a 95% Probability of Success

Organisms	Genome Size (g)	Colonies Screened (N)
M13 Phage	3.1×10^3	1
Escherichia coli	4.0×10^6	1840
Saccharomyces cerevisiae	1.35×10^7	6200
Drosophila melanogaster	1.8×10^8	82,200
Tobacco	1.6×10^9	735,998
Human	2.8×10^9	1,288,008
Zea mays (corn)	1.5×10^{10}	6,900,688

identified by agarose gel electrophoresis, subsequently excised, and then used as the DNA for the gene bank. Therefore, the researcher does have extensive control over which size fragments are isolated.

It is fundamentally important to look ahead to the screening of a library before preparing the DNA for cloning. Based on the organism, the number of clones screened will differ, and the amount of DNA needed for the library will change. Logically, as the number of clones requiring screening increases, so does the amount of DNA. However, usually even a small quantity of DNA (e.g., 5 µg) can be adequate for the entire library.

The number of clones screened is usually estimated prior to the experiment. Several factors are considered in these estimates, namely the probability for success (**P**), the size of the genome (**g**), and the size of the partial digest fragments (**f**). The formula used to estimate the number of clones is as follows:

$$N = \ln(1-P)/\ln(1-f/g)$$

N is the number of clones screened
P is the probability of isolating the gene
f is the fragment size
g is the size of the haploid genome

This formula provides an estimate (N) of the number of clones that should be examined to find a specific sequence. It also demonstrates that the researcher has control of two parameters, namely the probability of finding the gene (which is *never* 100%) and the size of the fragments.

As an example, suppose that you have a fungus with a haploid genome of 9×10^7 bases. Following a partial digest, you isolate fragments of 5000 bases. If you want a 95% chance of finding your sequence (assuming your sequence is smaller than 5000 bases), then the calculation would be as follows:

$$N = \ln(1{-}0.95)/\ln(1{-}(5{,}000/9 \times 10^7)).$$

In this, N is equal to 53,922 clones. This number represents the number of clones that should be examined in order to find one intact targeted sequence. By altering the parameters, the screening effort changes. For instance, increasing the probability of success to 99% increases the number of clones to 82,890. By increasing fragment size from 5,000 to 20,000 bases, for a probability of 95%, the number drops to 13,479.

This formula is simply an estimate, and it is important to note that the chance does exist that no clones will be found. Often extensive screening is necessary to isolate one clone.

The Screening Process

As noted in Chapter 7, nucleic acid probes are labelled oligonucleotides, interspecies genes, or cDNA that are homologous with the targeted DNA. The screening method using these probes is called blotting. Depending on both the type and specificity of a probe, the parameters for blotting will differ. However, blotting has many generalized features, including: (1) transfer and linking DNA to a membrane; (2) denaturing the DNA to single strands; (3) neutralization; (4) prehybridization of the membrane; (5) hybridization of a probe to the bound DNA; and (6) visualization of the probe.

To individually screen many thousands of clones is unrealistic. Accordingly, techniques for the mass screening of transformants have been developed; a common technique is the colony blot (or plaque lift if phage vectors are used). Colonies are grown on or transferred to a membrane, the DNA is released by lysing the cells on the membrane, and then the exposed DNA is analyzed with nucleic acid probes that hybridize to the target sequence. With plaque lifts, one petri dish may contain as many as 1000 different clones. Examining 100,000 transformants translates into screening 100 petri dishes, a much more realistic task than examining each transformant. When millions of colonies require screening, it is not uncommon for many members of a research team to be dedicated to the screening process.

With the colony blot, the first step is to grow or transfer colonies on to a membrane. *E. coli* are normally plated on selective agar following transformation, and the colonies that develop can be handled in several ways. Each individual colony can be picked and transferred to either an agar petri dish (containing a selective medium) or to a well of a 96 or 384 well microtiter plate. This picking and plating of the transformants yields an orderly library that can be repeatedly screened and cataloged. Alternatively, the petri dishes with transformants are indirectly screened via the less labor intensive replica plating.

Replica plating is a technique that involves lifting colonies from an agar surface onto a replica plating pad (i.e., a cloth, paper, or foam surface). That pad can then be used to imprint the original colonies onto new surfaces, including new agar plates and membranes (for blotting). Nylon and nitrocellulose membranes can be laid on an agarose surface, and colonies are imprinted on the surface of the membrane. The cells will grow since nutrients will seep through the membrane from the agar. One option is to imprint the colonies directly onto a membrane, i.e., colony lift.

Once transformants are lifted to or grown on a membrane, the cells must by lysed to release the DNA. This lysis is accomplished by laying the membrane (clone side up) on filter paper dampened with NaOH (0.5 N). The NaOH diffuses through the membrane causing cell lysis and the uncoiling (denaturation) of the DNA. The NaOH not only disrupts cells and denatures DNA, it also allows the DNA to adsorb to the membrane. Membranes used for blotting, such as nitrocellulose and nylon, are positively charged and thus bind negatively charged DNA. Following denaturation, the membrane is neutralized with a buffered salt solution, and then it is baked to fix the DNA to the surface. (Many times DNA is fixed by UV cross-linking. However, with colony blots, cell debris can sometimes interfere with cross-linking; therefore heat is used.) The membranes can now be probed.

Since DNA readily sticks to positively charged membranes, the entire membrane must be saturated with DNA and/or protein to prevent non-specific adsorption of the probe. The membrane is prehybridized by soaking it in a blocking solution which usually contains a sodium citrate/sodium chloride buffer (e.g., 5X SSC), casein (milk protein, 2%), N-laurylsarcosine (0.1%), and sodium dodecylsulfate (SDS, 0.1%). Many times sonicated, denatured salmon sperm DNA is added to the prehybridization solution, as well as formamide (denaturing agent), and dextran sulfate (water excluding agent). Membranes are blocked for at least 1–2 hours prior to adding the probe.

As noted, DNA released from colonies must be denatured before it is fixed to the membrane. Obviously, if the target DNA is double-stranded, then a probe cannot hybridize to its complementary sequence. Similarly, the probe itself must also be single-stranded, such as oligonucleotides are. However interspecies genomic and cDNA probes are double-stranded and require denaturation prior to use. This is accomplished by heating the probe (e.g., boiling for 10 min), followed by rapid cooling in an ice bath. Once denatured the probe is added directly to the prehybridization solution (the prehybridization solution now becomes the hybridization solution).

A physical parameter that greatly influences the specificity of the probe is the temperature of hybridization. Nucleic acids anneal/hybridize at any temperature below their T_m (i.e., temperature of melting). Two factors that affect T_m are probe base composition and length. Since guanine and cytosine form three hydrogen bonds during base pairing, but only two bonds between adenine and thymine, probes with higher G/C content have a higher T_m. Accordingly, the length of a probe will influence T_m as the number of hydrogen bonds change. A probe tends to anneal nonspecifically to a template at hybridization temperatures significantly lower than its T_m. This occurs when short regions of the probe hybridize to homologous sequences in the genomic DNA. To ensure the specificity of probes, hybridization temperatures near the T_m are normally used, e.g., T_m–5°C, with temperatures ranging from 42°C to 68°C often being employed.

Once a probe hybridizes to a target, it must be located. This detection is accomplished by initially labelling the probe so that it can be visualized following hybridization. Labelling involves incorporating a radioactive phosphorous or antigenic tag into the probe. The isotope of phosphorous, ^{32}P, is often incorporated into the phosphate backbone of nucleic acids. Once the ^{32}P labelled probe has hybridized, the blot is placed against X-ray film, and the decaying ^{32}P produces an image. Less hazardous techniques for detection involve directly or indirectly linking enzymes to the probe. These enzymes, such as alkaline phosphatase or horseradish peroxidase, are able to cleave special substrates which either turn color (colorimetric) or emit light (chemiluminescent). Such cold probes (cold is used to describe nonradioactive) can be just as sensitive as radioactive probes.

11.3   EXPERIMENTAL DESIGN AND PROCEDURES

In Chapter 7 we explored the options for the cloning of *MEL1* from *Saccharomyces carlsbergensis*. One of these options involved designing an oligonucleotide probe from the N-terminal amino acid sequence of the α-galactosidase. This set of laboratory experiments will focus on the labelling of an oligonucleotide probe and its application to find a clone. An alternative probe synthesis technique, random priming, is also presented. Following the identification of the clone, you will culture the organism in preparation for its analysis and characterization.

Tailing of an Oligonucleotide

When the only data available about a gene is the amino acid sequence of the protein, then oligonucleotides are used as the probes. Oligos (slang) are synthesized on DNA synthesizers, special instruments which can grow single-stranded DNA. Most oligonucleotides are less than 30 bases in length.

One method of labelling oligonucleotides is by adding nucleotides to the 3' end with terminal deoxytransferase (TdT), a process known as tailing. By combining the oligonucleotide, buffer, labelled deoxynucleotide (e.g., dig-11-dUTP), and enzyme, the label is tagged onto the probe. The label used in this manual is digoxigenin-dUTP, a derivative of dTTP. Digoxigenin is a naturally occurring plant steroid. Antibodies raised against digoxigenin have been conjugated to alkaline phosphatase. This antibody enzyme conjugate (linked) is used in an enzyme immunoassay to detect the probe after it hybridizes to its target.

Materials

Oligonucleotide tailing kit* or equivalent components

Oligonucleotide—Derived from the N-terminal amino acid sequence of α-galactosidase (e.g., 5'-TCACTCCACAGAT-GGGTTG-3')

3 M potassium acetate, pH 5.5

95% ethanol, ice cold

TE buffer—10 mM Tris, pH 8, 1 mM EDTA

Method

1. Mix the following reagents in a small microfuge tube. The volumes of oligonucleotide and water can be varied to obtain the correct concentration of DNA.

4.0	µl	5X Tailing Buffer
4.0	µl	$CoCl_2$ (25 mM)
__._	µl	Oligonucleotide (100 pmol)
1.0	µl	Digoxigenin-11-ddUTP (1 mM)
1.0	µl	Terminal Transferase (50 U/µl)
.	µl	Water
20.0	µl	Total

   (Tailing buffer (5X) contains 125 mM Tris-HCl, pH 6.6, 1 mM potassium cacodylate, 1.25 mg/ml BSA. Dig-11-dUTP can be used instead of Dig-11-ddUTP.)

2. Incubate the solution at 37°C for 15 min.

3. To stop the reaction, heat the tube to 65°C for 10 min. The oligonucleotide probe can be used directly, or cleaned and concentrated (see #4).

4. Add 2 µl of glycogen/EDTA solution (0.1 µg glycogen/ml in 200 mM EDTA, pH 8.0), then add $\frac{1}{10}$ volume potassium acetate and 2.5 vol-

---

*Genius™ Oligonucleotide Tailing Kit (Boehringer Mannheim Corp.)

umes of cold ethanol. Incubate at –20°C for 2 hr. Microfuge at high speed for 20 min. Decant, dry, and resuspend to a concentration of 10 pmol/µl in TE buffer. Store at –20°C until needed.

Synthesis of a Random Primed DNA Probe

Many types of probes are available for the detection of clones. As noted above, oligonucleotides are commonly used when only the amino acid sequence of the gene's protein is known. An alternative probe is a cDNA which can be used to probe for its genomic equivalent. There are numerous techniques for labelling probes, such as with tailing. One option which is used to label larger probes is random priming (see Chapter 7.2). In this experiment you will random prime label the *MEL1* gene.

Materials

Random primers*—10X random hexanucleotides

Deoxynucleotide triphosphates*—All four dNTPs in 10X concentration, included digoxigenin-11-dUTP.

Klenow fragment* (0.1 U/µl)

MEL1 template DNA—1 µg/µl

Sterile distilled or deionized water

Method

1. Mix the following components in a screw cap microfuge tube. As noted before, the volumes of DNA and water can be varied so to obtain the correct concentration of DNA.

_._	µl	distilled H_2O
2.0	µl	10X hexanucleotide mix
_._	µl	template DNA (10 ng to 3 µg)
17.0	µl	Total

*All components of the Genius™ Non-Radioactive Random Primed Labelling Kit

2. Cap the tube and denature the template DNA by boiling it for 10 min in a water bath. Heat blocks may not provide adequate temperature to efficiently denature the DNA. Transfer the DNA to an ice or preferably dry ice/ethanol bath until use. A dry ice/ethanol bath is simply made by placing dry ice in a Styrofoam insulated container and then covering the dry ice with ethanol. The ethanol will bubble and foam voraciously; however, after several minutes the bubbling will stop. **This bath is very cold. Take care not to burn yourself!** Dry ice/ethanol baths are very useful for flash freezing denatured DNA to prevent its reannealing. In synthesizing a random primed probe, this denaturation and rapid chilling/freezing of the solution are the most critical steps. If the DNA is not denatured, the random primers cannot anneal to the template.

3. Remove the tube from the ice bath and allow it to thaw to room temperature (this will occur rapidly). Add 2 µl of 10X dNTP labelling mix and 1 µl of Klenow fragment. Mix and incubate the solution at 37°C for 1 to 20 hrs. If necessary, centrifuge the tube to pellet condensation. The longer the incubation, the more digoxigenin will be incorporated into the DNA probe.

4. Following the incubation, the probe should be tested for the incorporation of the label (see below). The labelled DNA can be used immediately or stored frozen (–20°C) for up to a year.

Testing of Digoxigenin Labelled Probes

Each step of a research process needs verification, and probes are no exception. The extent by which the label was incorporated into the probe during its synthesis should be assessed before the probe is used in the colony blot.

Materials

Nylon membrane—3 cm × 4 cm
Clean petri dish
Random primed probe or tailed oligonucleotide

MORE...

Control labelled DNA or oligonucleotide*

Blocking solution—2% Blocking Powder* or BSA in 100 mM Tris, pH 7.5, 150 mM NaCl

Anti-digoxigenin alkaline phosphatase conjugate*

Wash solution—100 mM Tris, pH 7.5, 150 mM NaCl

Substrate buffer—100 mM Tris, pH 9.5, 150 mM NaCl, 50 mM MgCl$_2$

X-Phos and NBT stocks*

Method

1. Separately prepare a serial dilution of both the probe and control DNA of 10^{-1}, 10^{-2}, 10^{-3}, 10^{-4}, and 10^{-5} in water.

2. In two rows (one row for the control and the other for the probe), in series spot 1 µl of the dilutions onto the membrane. Use a pencil to indicate where the dilutions were spotted (do not mark the exact spot).

3. Place the membrane face down on a UV transilluminator and cross-link the DNA to the membrane for 3 min. This cross-linking covalently binds the DNA to the membrane. **UV light can cause damage to your skin and eyes. Wear UV safe goggles and avoid excessive exposure to UV light.**

4. In the petri dish, soak the membrane for 15 min in 10 ml of blocking solution.

5. Add 2 µl of anti-digoxigenin alkaline phosphatase conjugate to the blocking solution. Incubate for an additional 15 min.

6. Rinse the membrane twice for 5 min with wash solution. This wash removes unbound antibody.

7. Soak the membrane for 1 min in substrate buffer.

8. Prepare substrate solution by adding 45 µl NBT (nitroblue tetrazolium at 100 mg/ml in DMF) and 35 µl X-Phos (5-bromo-4-chloro-3-

---

*Indicated materials are provided with the Genius™ labelling and detection kits.

indoyl-phosphate at 50 mg/ml in DMF) into 10 ml of substrate buffer (100 mM Tris, pH 9.5, 150 mM NaCl, 50 mM MgCl$_2$). This substrate is light sensitive, so avoid prolonged exposure to the light. Excess solution may be frozen if desired.

9. Replace the substrate buffer with 10 ml of substrate solution. A color precipitate should develop within 15 min.

10. By visually comparing the probe to the diluted control DNA, estimate the concentration of the probe (Figure 11.1).

Colony Lifting and Probing of Transformed *E. coli*

The cloning of DNA uses either phage or plasmid vectors, of which phage vectors are normally used for genomic cloning. This is because lambda

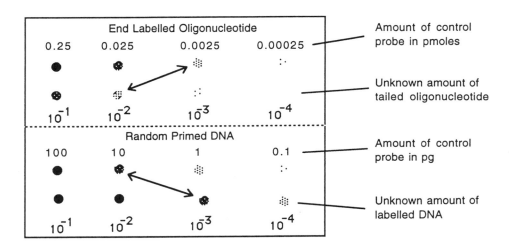

Figure 11.1 Schematic of estimating probe concentration by membrane spotting. Known standards are serially diluted and compared to serially diluted labelled probes. Upon probe visualization, color intensities between the known concentrations and unknown probes are compared and matched (indicated by arrows). After factoring in the dilutions, the oligonucleotide is assumed to have a concentration of 0.25 pmol/μl while the random primed DNA has a concentration of 10,000 pg/μl or 10 ng/μl.

phage vectors can accommodate very large inserts and also because screening recombinant phage libraries is more efficient than screening colonies. Plasmid based libraries, however, are easier to construct and introduce into *E. coli*. Due to logistics, we will be screening a plasmid based library.

Materials

Petri plates with transformants—refrigerated at 4°C

Nylon membranes—circular to fit a petri dish

Denaturing solution—0.5 N NaOH, 1.5 M NaCl

Neutralizing solution—2 M NaCl, 1 M Tris, pH 7.5

Prehybridization solution—5X SSC, 1.0% Blocking Reagent,* 0.1% N-lauroylsarcosine, 0.02% sodium dodecyl sulfate. 20X SSC is 0.3 M Na citrate, pH 7, 3 M NaCl.

Heat sealable plastic bags (hybridization bags)

Heat/impulse sealer

Digoxigenin labelled probe

Method

1. Mark a nylon membrane at the edge with a pencil. (Gloves should be worn while handling the nylon membrane to avoid contaminating the membrane.) Also mark the bottom of the petri dish at the edge as well. Gently place the nylon membrane on the agar so that the marks on the membrane and petri dish align. This aligning of the membrane and dish is critical since later you must use the membrane as a guide to find positive clones. Allow the membrane to sit for 5 min. Take extra care not to smear or flatten the colonies on the plate during this transfer. While the cells are transferring, prepare two small trays (slightly larger than the nylon membrane) with several pieces of filter paper on each. On the first tray saturate the

*Blocking Reagent is a 10% solution of Blocking Powder (Boehringer Mannheim Biochemicals) in 100 mM Tris, pH 7.5, 150 mM NaCl. First combine the Tris and NaCl, and then add the Blocking Powder. Warm the solution to dissolve the Blocking Powder. Do Not Boil. Excess Blocking Reagent can be stored frozen to prevent spoilage.

filter papers with denaturing solution and the second tray with neutralizing solution. The membrane will be placed colony side up onto these filters. Excessively damp filters may cause the colonies to run and thus smear any positive hybridization signals. Consequently, it is important to adjust the wetness of the filters in order to avoid this problem. Excessive solutions can be drawn off with paper towels or extra filters.

2. Carefully peel the membrane off the agar surface with forceps and place *colony side up* on the filter paper moistened with denaturing solution. Incubate for 15 min at room temperature.

3. Transfer the membrane to dry filter paper for five minutes. The membrane is then transferred to the filter papers moistened with neutralizing solution and is incubated for 15 min.

4. Again the membrane is blotted on a filter and placed in a 80°C oven and baked to fix the DNA to the surface of the filter. Since we are using nylon membranes, a vacuum oven is not necessary; however, if nitrocellulose membranes were being used, then a vacuum oven would be necessary since 80°C is above the flash point of nitrocellulose. Due to cellular debris, linking the DNA to the filter with UV light (an option in the blotting protocols) will not be necessarily successful.

5. Place the membrane in a heat sealable bag. Add 10 ml prehybridization solution to the bag, expel the bubbles, and heat seal. The next step is to prehybridize the membrane with a protein solution to block against nonspecific adsorption of the probe. The temperature of prehybridization is dependent upon the type of probe that is to be used. Prehybridize the membrane for 1–2 hr at 68°C for random primed probes and T_d–5°C for oligonucleotides. The temperature at which an oligonucleotide dissociates from its homologous sequence is T_d. This is calculated from the nucleotide content of the probe as follows:

$$T_d = [4 \times (G + C)] + [2 \times (A + T)]$$

To ensure that the probe hybridizes to its specific sequence, a hybridization temperature of T_d–5°C is used.

6. Following the incubation, the prehybridization solution should be poured into a clean tube and the membranes wiped with a prehybridization solution-dampened clean cloth (Kimwipe) so that any cellular debris is removed. Do not forget to wear gloves. This handling of the membrane would be strictly avoided if the detection method employed chemiluminescence. Replace the membrane into the bag.

7. If a double-stranded probe is being used (e.g., random primed labelled DNA), denature the probe by heating in a boiling water bath for 10 min and immediately immersing the probe in an ice bath or dry ice/ethanol bath. Oligonucleotide probes do not require denaturing since they are single-stranded. Mix the probe directly into the prehybridization solution to the following concentration:

Oligonucleotide Probe   1–10 pmol/ml

DNA Probe   5–20 ng/ml

The amount of probe used will be dependent upon its concentration as determined in the previous experiment. The remaining probe should be placed at –20°C since it can be used again in Chapter 12.

8. Add the probe solution (hybridization solution) to the bag. Expel all bubbles and seal the bag. Incubate at 68°C overnight for random primed probes or at T_d–5°C for one to six hours for oligonucleotide probes. Following the incubation, continue on to the visualization step.

Visualization of Colony Blots

Two techniques are predominantly used for the visualization of nonradioactive probes, namely colorimetric and chemiluminesence, with the latter being more sensitive. In colony blotting, however, background caused by cellular debris makes chemiluminescence unsuitable. Alternatively, colorimetric detection is sufficiently sensitive for colony blotting due to the high concentration of the cloned DNA in each colony,

Visualizing your probe is accomplished by an enzyme immunoassay. An anti-digoxigenin antibody which is linked (conjugated) to alkaline phosphatase (i.e., anti-dig alkaline phosphatase conjugate) will bind to the probe/target DNA hybrid. Alkaline phosphatase can then hydrolyze the substrate X-Phos/NBT, which causes a blue precipitate to form at the site.

Materials

2X SSC, 0.1% SDS solution

0.5X SSC, 0.1% SDS solution at 65°C

Blocking solution—2% Blocking Powder* or BSA in 100 mM Tris, pH 7.5, 150 mM NaCl

Wash Solution—100 mM Tris, pH 7.5, 150 mM NaCl

Substrate buffer—100 mM Tris, pH 9.5, 150 mM NaCl, 50 mM MgCl$_2$

Substrate solution—X-Phos and NBT* in substrate buffer

Anti-digoxigenin antibody alkaline phosphatase conjugate*

Method

1. Retrieve your colony blot from the water bath. Open the bag and pour off the hybridization solution. This solution can be saved (frozen) and reused several times (up to five times for a DNA probe). Hybridization solutions that contain double-stranded probes must first be boiled to denature the probe before reuse. Wash your membrane for 15 min with 2X SSC, 0.1% SDS in the bag at room temperature. Repeat the washing.

2. At 68°C for random primed probes or T_d–5°C for oligonucleotides, wash the membrane in 0.5X SSC, 0.1% SDS for 15 min. Washing at elevated temperatures can be done by sealing preheated 0.5X SSC and membrane in a plastic bag and floating the bag in a water bath. Pour off the solution. Repeat this washing step.

3. Soak the membrane in wash solution for 1 min at room temperature. Pour off the solution.

4. Add 10 ml of blocking solution to the bag and incubate for 30–60 min at room temperature with shaking or occasional agitation. This step

*Items supplied with Genius™ kits.

can go overnight if necessary. The blocking solution essentially covers all free and open areas of the membrane with protein. This prevents nonspecific adsorption of the antibody–conjugate to the membrane.

5. Dilute the antibody–enzyme conjugate 1:5000 in blocking solution (2 µl). The antibody can be added directly to the blocking solution in the bag; however, don't touch the membrane when introducing the antibody. Incubate for 30–60 min at room temperature.

6. In the bag, wash with wash solution twice for 15 min. Pour off.

7. Add 10 ml of substrate buffer and allow the membrane to soak for 2 min. Pour off the solution and add 10 ml of substrate solution. Seal the bag and incubate in the dark for as little as 15 minutes and up to 12 hours. Colonies that bind the probe will be dark purple. Be careful not to mistake negative colonies which may react faintly as compared to the positive clones.

8. Wash the membrane with wash solution to stop the reaction. The blot can then be dried and stored in a sealed bag. It is best to photocopy or photograph the blot shortly after visualization since the color typically fades with time (i.e., photobleaches).

Picking and Culturing of Transformed Colonies

In molecular biology, the term clone can refer to the DNA molecule you have constructed in vitro. When introduced into a host, the plasmid replicates at up to 200 copies per cell, with each molecule being an exact duplicate of the original. It is assumed that each colony arose from one plasmid construct assimilated by a single cell. Thus, when you identify a positive colony, you have isolated an individual clone. What remains is the identification, verification, and characterization of the clone. This may be as simple as screening for an observable phenotype, or it may require intricate and extensive characterization (i.e., antibody reactions with the protein product, hybridization, or DNA sequencing). In this exercise you will identify and culture the clone in preparation for characterization.

Materials

        Colony blot
        Plates with transformants
        Sterile toothpicks
        Tubes with 5 ml sterile LB/ampicillin broth (50 µg/ml ampicillin)

Method

1. Place your blot colony side up on a counter. Place the original agar plate over the blot, lid side down. Turn the plate so that the marks on the membrane and plate are aligned. Using a marker, circle any colonies on the plate that appear as positives on the blot.

2. Carefully pick the positive colonies from your plate with a sterile toothpick. Place the tip of the toothpick in the opening of the culture tube (containing the broth) and break the toothpick into the tube. Breaking the toothpick is done by bending it against the lip of the tube. Don't touch the lower part of the toothpick. It must remain sterile. In practice, store the clone on a plate (record the phenotype for each colony transferred) so that you can return to a particular colony once you have determined that it is your clone.

3. Incubate culture tubes at 37°C overnight with agitation. The cells can be frozen until needed.

4. Using the Plasmid DNA Isolation protocol (Chapter 8.3), isolate the plasmid.

5. Quantitate the DNA isolated using UV spectroscopy (Chapter 8.3).

6. Store the plasmid solution at –20°C until needed.

STUDY QUESTIONS

1. What is the advantage of using a random primed probe over an oligonucleotide probe?

2. Following a colony lift and probing, what difficulty could be experienced when trying to identify the original colony? What could you try to overcome the problems?

3. Following the synthesis of a random primed probe, testing reveals a very poor yield of probe. What could be responsible for the low yield?

FURTHER READINGS

Berger SL, Kimmel AR (1987): *Guide to Molecular Cloning Techniques*. San Diego: Academic Press

Old R, Primrose S (1989): *Principles of Gene Manipulation*. Oxford: Blackwell Scientific Publications

Sambrook J, Fritsch EF, Maniatis T (1989): *Molecular Cloning: A Laboratory Manual*. Plainview, NY: Cold Spring Harbor Laboratory Press

12

Characterizing and Verifying Cloned DNA

12.1   OVERVIEW

It is necessary to clone DNA prior to its molecular analysis and subsequent manipulation and application. Once cloned, its characteristics such as fragment size, restriction sites, nucleotide sequence, and subsequence identification (biologically significant sequences) are forthcoming.

Clones are usually between 7–20 Kb, but the targeted sequence (i.e., the gene) is usually only a portion of that fragment. Gross characteristics, such as fragment size and restriction endonuclease sites, can be deduced by combining restriction digestion with agarose gel electrophoresis (Chapter 9) in a technique called restriction mapping. However, restriction mapping does not locate the target sequence. Combining restriction mapping with Southern blotting will determine the location of the target sequence on the clone. Determining the size of the gene (located within the fragment), the precise location of restriction sites, and the identification of subsequences (e.g., start codons, promoters, introns), require detailed data which is obtained by DNA sequencing.

This technique combines DNA polymerization (similar to that used in a random probe synthesis) with polyacrylamide gel electrophoresis.

In the previous laboratory experiments, you performed the steps needed to isolate DNA from yeast, fragment it, ligate the fragments to a vector, insert the recombinant molecules into *E. coli*, and then screen those transformed *E. coli* to locate a clone with the *MEL*1 gene. The next logical step in this cloning scheme is to characterize the isolated clone. The following laboratory exercises will focus on restriction mapping, Southern blotting, and techniques used for DNA sequencing. The analysis of the DNA sequence will be examined in Chapter 13.

12.2   BACKGROUND

Cloning is the process by which DNA is isolated, manipulated, and biologically produced so that it can be subsequently analyzed and/or exploited. Analysis can take many forms, the most common being restriction mapping, Southern blotting, and DNA sequencing. Data generated from these techniques may reveal important chemical and biological features about the DNA and/or may provide information that can be used for its manipulation.

Restriction Endonuclease Mapping

Restriction mapping is an analytical technique used to assess whether and where restriction endonucleases cleave within DNA. By cleaving the vector/insert with a battery of restriction enzymes coupled to fragment analysis, the relative location of sites can be estimated based on the fragments produced. Individual fragments can then be tested for the presence of additional restriction endonuclease sites.

When a clone is isolated, an initial objective is to measure its size. Since the clone was generated from a restriction digest of genomic DNA and ligation into a vector, cleaving the recombinant vector/clone with the original enzyme liberates the insert from the vector. The fragment is then analyzed by gel electrophoresis against known standards in order to determine its size. If the ligation of the clone and vector destroyed the restriction site (e.g., as with ligating a *Nde*II fragment into a *Bam*HI site), then the fragment cannot be easily excised. The size of the insert, however, can be determined indirectly by linearizing the recombinant plasmid (i.e., cutting it at a unique site), electrophoresing, and then measuring the size. The insert size is the difference between the vector and the vector plus insert. That is:

Recombinant Plasmid (Kb) – Vector (Kb) = Insert (Kb).

The actual mapping of restriction sites is much more involved than determining the size of a clone. Restriction mapping, as it is called, involves establishing which restriction endonucleases cut within the DNA and establishing the relative position of their sites. Assessing which enzymes cleave DNA is a simple exercise in screening. DNA is cleaved with different enzymes and then analyzed by gel electrophoresis. Deviations in mobility of the digest as compared to an uncut control indicate the presence of at least one restriction site. When circular DNA molecules are cleaved, the number of bands generated is equal to the number of restriction sites; thus one band indicates one site, two bands for two sites, and so on.

After it is cleaved, a plasmid with a unique restriction site will appear as one band on a gel. If the digest is repeated with the same plasmid containing an insert, then one band on a gel would indicate that there are none of those restriction sites in the insert. Two bands would indicate that there is a site in the plasmid (as expected) and one in the insert. Also, the size of the fragments can be used to deduce the distance between and positions of the restriction sites.

The logic used in deducing the relative location of restriction sites is difficult to describe. It is much like piecing together a jigsaw puzzle. The fragments are often rearranged on paper until the pieces fit. Often it can be very difficult to account for all the information generated. As an example of restriction mapping, the following hypothetical scenario is presented.

A gene bank is synthesized by ligating *Nde*II fragments from a donor into a *Bam*HI digested vector (e.g., pUC119). Screening the library yields a clone which is subsequently analyzed. The restriction map of the plasmid is shown in Figure 12.1.

Ligation of a *Nde*II fragment into the vector's *Bam*HI site destroys that site. Cutting the plasmid with a different unique restriction endonuclease, such as *Eco*RI, could yield the pattern shown in Figure 12.2.

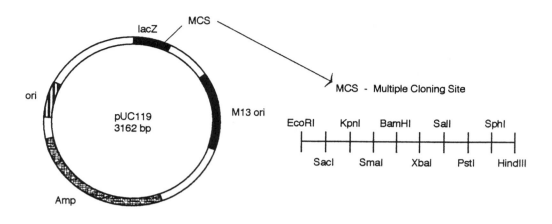

Figure 12.1 The vector pUC119. The MCS (multiple cloning site or polylinker) contains many unique restriction endonuclease sites that are available for cloning.

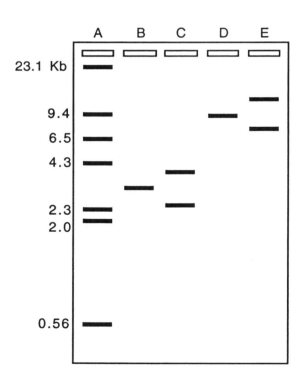

Figure 12.2 Electrophoretic pattern of *Eco*RI digested vector/insert. Lane A—*Hind*III digested lambda DNA, Lane B—*Eco*RI digested pUC119, Lane C—undigested pUC119, Lane D—*Eco*RI digested pUC119/Insert, Lane E—undigested pUC119/Insert.

This electrophoretic map yields certain information. Since one band appeared in the *EcoRI* digested pUC119/Insert lane, there is no *EcoRI* site in the Insert. Also, by plotting the migration of the lambda fragments against their mass, a standard curve can be established. The length of the plasmid is 3100 bases, while measuring the cut recombinant plasmid suggests a size of 9000 bases. By subtraction, the Insert is 5900 bases (9000 − 3100 = 5900). From this, a simple restriction map can be constructed (Figure 12.3).

The mapping is continued by digesting the recombinant plasmid with a battery of restriction endonucleases. For argument, let the enzymes *HindIII* and *PstI* generate the following electrophoretic map (Figure 12.4).

The most obvious conclusion from this data is that *HindIII* cleaves the recombinant plasmid twice while *PstI* cleaves in three locations. When compared to lambda fragments, the *HindIII* bands are 1300 and 7700 bases while the three *PstI* fragments are 700, 2300, and 6000 bases. This data can be used to update the restriction map (Figure 12.5). All but one of the restriction endonuclease sites are easily plotted.

The rationale for designing this map is as follows:

(1) The small *HindIII* fragment is generated from the site on the plasmid and a site 1300 bases into the insert. This leaves 4600 bases of insert plus 3100 bases of the vector (4600 + 3100 = 7700).

(2) Of three *PstI* fragments, only the 6000 base fragment is sufficiently large to contain the plasmid. Subtracting the length of the plasmid

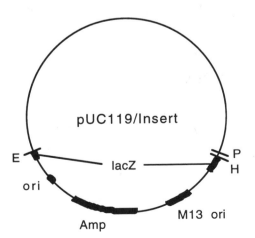

Figure 12.3 Updated restriction map of pUC119/Insert. E = *EcoRI*, P = *PstI*, and H = *HindIII*.

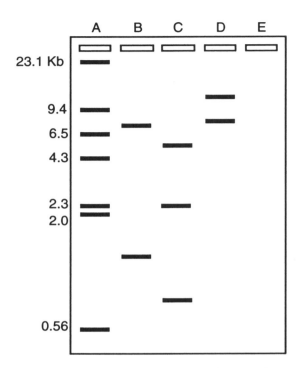

Figure 12.4 Electrophoretic map of pUC119/Insert cleaved with *Pst*I and *Hind*III. Lane A—*Hind*III digested lambda DNA, Lane B—*Hind*III digested pUC119/Insert, Lane C—*Pst*I digested pUC119/Insert, Lane D—undigested pUC119/Insert, Lane E—empty.

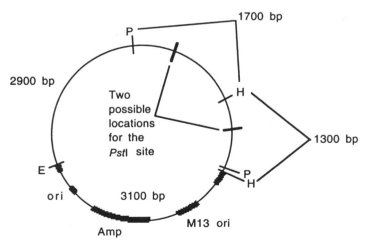

Figure 12.5 Updated pUC119/Insert restriction map including sites for *Eco*RI, *Hind*III, and *Pst*I.

(3100 bases) indicates that the distance from the original plasmid *Bam*HI site to the Insert *Pst*I site is 2900 bases. The location of the second Insert *Pst*I site is impossible to precisely determine. The site can be positioned 700 bases from either *Pst*I site, which would generate the same two 2300 and 700 base fragments.

(3) To determine the relative location of this unknown *Pst*I site requires digesting the plasmid with two different restriction endonucleases simultaneously, namely *Pst*I and *Hind*III. The question which remains is where in the 3000 base fragment does a *Pst*I site reside (Figure 12.6).

The fragment under question is defined by two *Pst*I sites and characterized by the *Hind*III site. The arrows are the two possible choices for the unknown *Pst*I site. Cutting this fragment at either location would yield the 700 and 2300 base fragments. The asymmetrically located *Hind*III site is the key to solving this puzzle. If the unknown *Pst*I site was at "A" (Figure 12.6), then cutting the fragment with both *Pst*I and *Hind*III would generate Figure 12.7.

Thus the generation of 700, 1000, and 1300 base fragments on a gel would support the *Pst*I site being location at "A." If "B" was the location of the *Pst*I site (Figure 12.6), then the following would be true (Figure 12.8).

This progression of experiments and deductions is typical of the steps involved in restriction mapping. The above example is not overly complex, but illustrative of the process. At times, deciphering the data produced from restriction mapping can be quite complex, especially when an enzyme cleaves many times.

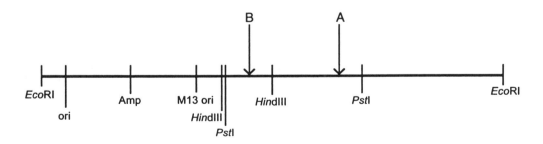

Figure 12.6 Possible locations for an unknown *Pst*I site.

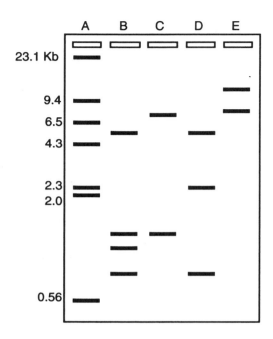

Figure 12.7 One possible restriction pattern after cutting pUC119/Insert with *Pst*I and *Hind*III. Lane A—*Hind*III digested lambda DNA, Lane B—*Hind*III and *Pst*I digested pUC119/Insert, Lane C—*Hind*III digested pUC119/Insert, Lane D—*Pst*I digested pUC119/Insert, Lane E—undigested pUC119/Insert.

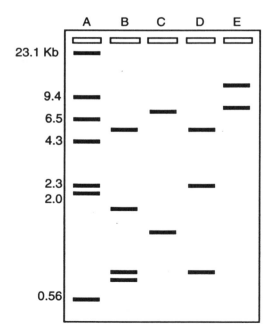

Figure 12.8 Electrophoretic map of alternative location of *Pst*I site. Lane A—*Hind*III digested lambda DNA, Lane B—*Hind*III and *Pst*I digested pUC119/Insert, Lane C—*Hind*III digested pUC119/Insert, Lane D—*Pst*I digested pUC119/Insert, Lane E—undigested pUC119/Insert.

Southern Blotting

The fragments generated for genomic cloning can be proportionally large in comparison to the genes they contain. Fragment sizes typically range from 7–20 Kb while a complete gene (without introns) may be 1 to 3 Kb. In some instances, such as the Human Genome Project (i.e., the sequencing of the entire human genome), special clones have been constructed which are 1–2 *megabases*. Accordingly, clones may include a large amount of sequence which is not immediately useful. Restriction mapping elucidates the positions of the restriction endonuclease sites, but it in no way identifies the location of the desired sequence. By combining restriction mapping with a variation of the colony blot, restriction fragments that contain the specific sequences can be identified. This process is a Southern blot, named after E.M. Southern who developed the technique in the 1970s.

Southern blotting involves several steps, which include: (1) cutting purified DNA (i.e., the clone) with a restriction endonuclease; (2) separating the fragments by gel electrophoresis; (3) transferring (blotting) the DNA from the gel to a membrane (nitrocellulose or nylon); (4) hybridizing a nucleic acid probe with the target; and (5) visualizing the probe.

Using this technique, it is possible to both verify (or detect) a DNA sequence and locate the fragment in which the homologous sequence resides.

Digesting DNA with a restriction endonuclease and electrophoretically separating the fragments has been described previously (see Chapter 8). Following electrophoresis, the gel is stained and photographed. Denaturation of the DNA to single-strands is accomplished by soaking the gel in a NaOH/NaCl solution which separates the double helix. This denaturation is extremely important as probes can not hybridize to their homologous sequences unless both are single-stranded. The gel is subsequently neutralized in a neutral pH, high salt buffer.

DNA in the gel must be transferred to a nitrocellulose or nylon membrane. Several methods are used for this transfer, namely capillary blotting, electroblotting, or vacuum blotting. Capillary is the original and most common means of blotting. The basis for the technique involves drawing buffer and DNA from the gel through the membrane by capillary action. By sandwiching the membrane between the gel and absorbent paper, the DNA will be pulled onto the membrane as buffer flows

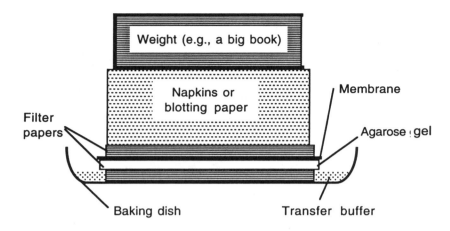

Figure 12.9   Southern blotting by capillary action.

into the absorbent paper from the gel. The gel is placed on a stack of filter paper in a dish. A buffer (neutral pH and high salt) is added to the dish as a mobile phase to aid in the DNA transfer (Figure 12.9).

DNA adsorbs to nylon and nitrocellulose due to positive charges on the membrane surface. To avoid deadsorption, the DNA must be covalently linked to the membrane. This is accomplished by heat or UV fixation.

To avoid nonspecific adsorption of the probe to the membrane, the surface must be blocked prior to hybridization. This prehybridization step, the probe hybridization, and the visualization are analogous to colony blotting. Probes that hybridize to the blot are detected either directly, as with radioactive probes, or indirectly through an enzyme-linked assay. In both instances, Southern blots are visualized on film with a positive result appearing as a black band (Figure 12.10).

DNA Sequencing

Restriction mapping and Southern blotting together are useful in locating a gene within a larger fragment. However, the elucidation of the actual nucleotide sequence requires DNA sequencing, a precise technique equivalent to the primary sequencing of a protein. The predominant method used to sequence DNA is Sanger dideoxy chain termination which combines DNA polymerization with acrylamide gel electrophoresis. The steps involved in the chain termination reaction are as follows:

(1) An oligonucleotide primer homologous to the template DNA is either purchased or synthesized. This primer, template, and polymerase buffer are combined and heated to a temperature that ensures all DNA is denatured.

(2) The denatured DNA solution is allowed to cool so that the primer anneals to its homologous sequence. This primer template solution is then divided into four tubes, each of which are labelled A, T, G, and C.

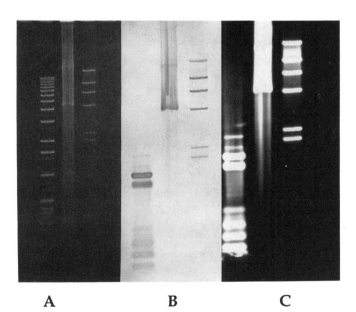

A   B   C

Figure 12.10 A comparison of DNA separated on an agarose gel and its corresponding Southern blot. (A) The agarose gel with ethidium bromide-stained DNA. (B) Southern blot with colorimetric detection. (C) Southern blot with chemiluminescent detection. Note the greater sensitivity of chemiluminescent detection as indicated by more numerous and darker bands.

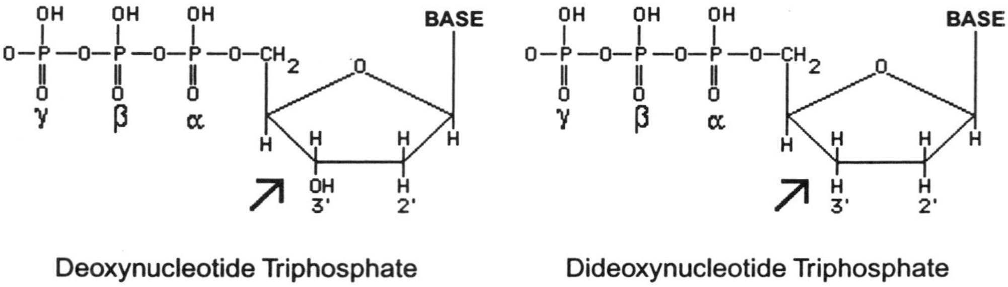

OH   OH   OH   BASE   OH   OH   OH   BASE

O—P—O—P—O—P—O—CH$_2$ O—P—O—P—O—P—O—CH$_2$

 γ   β   α   γ   β   α

OH   H   H   H
3'   2'   3'   2'

Deoxynucleotide Triphosphate   Dideoxynucleotide Triphosphate

Figure 12.11   A comparison between a deoxynucleotide and dideoxynucleotide.

(3) To each of the labelled tubes, a solution of dNTPs and a correspond-
     ing dideoxynucleotide triphosphate (ddNTP) is added. These
     ddNTPs are analogs of deoxynucleotides, the difference being that
     the hydroxyl group on the 3' carbon is replaced with a hydrogen
     (Figure 12.11). Though polymerases do prefer to incorporate dNTPs
     into a growing DNA strand, ddNTPs can also act as a substrate for
     DNA polymerases. Consequently when a ddNTP is incorporated
     into a growing DNA strand, the strand is terminated (stopped) since
     the 3' end lacks a hydroxyl (which is needed for the addition of the
     next nucleotide).

The preparation of four separate termination reactions, i.e., separate
tubes with ddATP, ddTTP, ddGTP, and ddCTP, yields a population of
terminated chains at every nucleotide along the template. For instance,
consider the following template (complementary strand):

3' – A C G C T T A G C G G A T G C A A T C C G T G – 5'

The first aspect of dideoxy sequencing is to denature the template
and anneal a primer, the result which could be as follows:

5' – T G C G A – 3'

3' – A C G C T T A G C G G A T G C A A T C C G T G – 5'

Next a ddNTP/dNTP solution and polymerase are added. It is important to note that the dNTP solution contains all four deoxynucleotide triphosphates. For this example, let the reaction contain ddTTP. The terminated products of this reaction are as follows:

$5'-TGCGAA_{dd}T$

$5'-TGCGAA\ TCGCC_{dd}T$

$5'-TGCGAA\ TCGCC\ TACG_{dd}T$

$5'-TGCGAA\ TCGCC\ TACG\ T_{dd}T$

$3'-ACGCTT\ AGCGG\ ATGC\ A\ ATCCGTG-5'$

These termination products can be separated and resolved by polyacrylamide gel electrophoresis. This separation is accomplished on specialized DNA sequencing electrophoresis units which are exceptionally tall, i.e, 40 cm, so that each termination product can adequately separate from the next. Using a gel with 6% acrylamide, single-stranded termination products that differ in mass by one base (i.e., 330 daltons) can be clearly separated.

As noted earlier, sequencing by the Sanger dideoxy method requires four different termination reactions, one for each base. These four termination reactions represent a pool of every possible termination product, i.e., one corresponding to each base in the sequencing template. When the termination products from the four reactions are resolved side by side, an electrophoretic ladder is produced which corresponds to the sequence of the DNA (Figure 12.12). Termination products that migrate the farthest represent bases closest to the primer while unresolved bands at the top of the gel are very long products.

Unlike DNA in an agarose gel, DNA in a sequencing gel cannot be visualized using ethidium bromide and a UV transilluminator. The mass of DNA used in sequencing is relatively small compared to the amount used in standard agarose gel electrophoresis; consequently alternative visualization methods are used. By far the most common detection methods employ radioactive nucleotides, such as ^{35}S labelled dATP. A relatively new isotope used for sequencing is ^{33}P. The strong signals produced by ^{32}P usually cause loss in resolution so ^{32}P is not normally used for sequencing reactions.

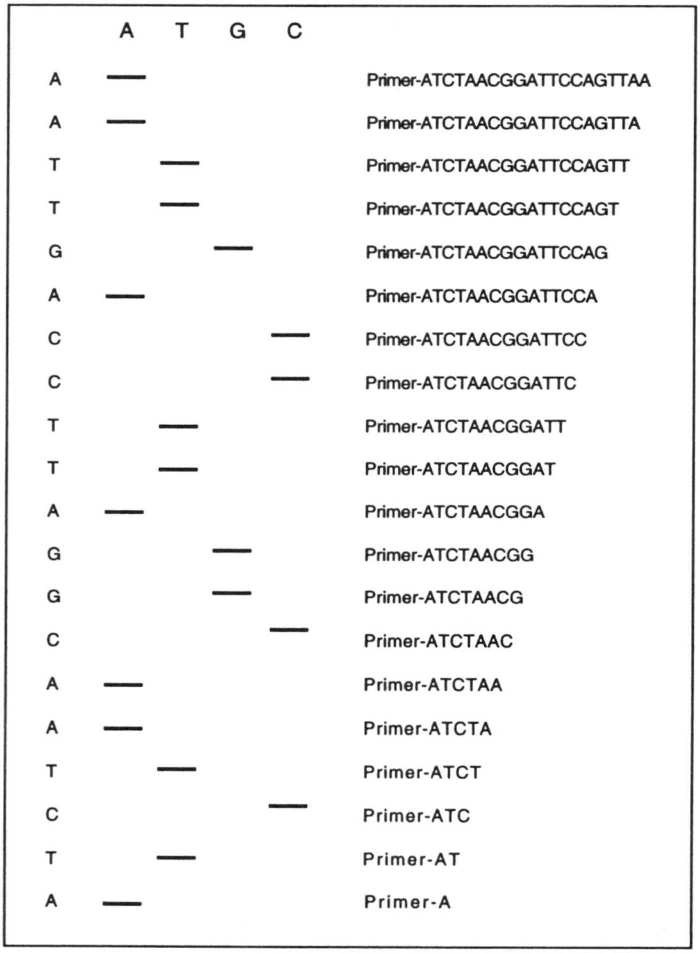

	A	T	G	C	
A	—				Primer-ATCTAACGGATTCCAGTTAA
A	—				Primer-ATCTAACGGATTCCAGTTA
T		—			Primer-ATCTAACGGATTCCAGTT
T		—			Primer-ATCTAACGGATTCCAGT
G			—		Primer-ATCTAACGGATTCCAG
A	—				Primer-ATCTAACGGATTCCA
C				—	Primer-ATCTAACGGATTCC
C				—	Primer-ATCTAACGGATTC
T		—			Primer-ATCTAACGGATT
T		—			Primer-ATCTAACGGAT
A	—				Primer-ATCTAACGGA
G			—		Primer-ATCTAACGG
G			—		Primer-ATCTAACG
C				—	Primer-ATCTAAC
A	—				Primer-ATCTAA
A	—				Primer-ATCTA
T		—			Primer-ATCT
C				—	Primer-ATC
T		—			Primer-AT
A	—				Primer-A

Figure 12.12 Ladder of bases produced from a sequencing gel.

Visualizing a sequencing gel involves the following:

(1) Label the primer with a radioactive or hot nucleotide. Hot nucleotides can also be incorporated into the termination products by simply adding a labelled nucleotide to the termination solution.

(2) Perform the sequencing reaction and separate the products on a sequencing gel.

(3) Soak the gel in a 10% methanol, 10% acetic acid solution to remove urea which will interfere with gel drying. Remove the water from the gel, i.e., dry the gel, on a gel drier, a type of vacuum manifold. Drying the gel concentrates the termination products into a thin cross section and prevents their diffusion in the gel.

(4) Wrap the gel in a plastic wrap (i.e., sandwich wrap) and place it against X-ray film in a X-ray cassette. Expose the film for 24 to 48 hours.

(5) Develop the film manually or in an automated developer.

In addition to the traditional methods, nonradioactive alternatives have also been developed. 5'-Digoxigenin or biotin labelled primers are used similarly to radioactive primers in the sequencing reaction and in the final visualization steps. However, additional steps in the middle include the transfer of DNA from the sequencing gel to a membrane, blocking the membrane, adding an anti-digoxigenin alkaline phosphatase antibody, followed by chemiluminescent detection. A different nonradioactive sequencing method involves incorporating differently colored fluorescent markers into each termination reaction. The individual bands in each lane can then be detected based on the color of light emitted when excited by UV light. This type of fluorescent label is used in automated DNA sequencing instruments.

Once a DNA sequence has been elucidated, computer analysis of the sequence is common. The sequence of nucleotides in DNA possesses all information necessary for life. This information is not only found in the coding regions of genes, but it is also found in promoters, terminators, enhancers, telomeres, origins of replication, centromeres, protein binding sites, introns, and important sequences yet to be discovered. Sequence analysis software is available that can rapidly and accurately dissect a DNA sequence. The importance of DNA sequence data and potential applications will be highlighted in Chapter 13.

Verifying the Clone

When a probe hybridizes to a colony or plaque, it can only be assumed that the clone is truly the DNA of interest. Often false positives occur when the probe inadvertently hybridizes to the wrong template. To differentiate between true positives and false positives additional analysis of the clone is necessary.

Sequencing provides information that can be used to verify the clone. For instance, if the primary amino acid sequence of the protein is known, then a corresponding DNA sequence indicates the clone is genuine, i.e., one in which the codons of the gene match the amino acids in the protein. The difficulty of verifying through sequencing is that not all protein sequences are known. Indeed, often genes are cloned to produce adequate amounts of protein for analysis. Also, it is common to do limited protein sequencing to provide sufficient data for probe design, but the entire amino acid sequence is not determined.

One approach to verifying a clone relies on the concept of cloning by complementation. If a gene is knocked out in a host and a loss in phenotype is observed, then introducing the clone back into the host will restore the phenotype and, it is hoped, indicate the clone is genuine. Reversion to the original phenotype is a strong indication of clone validity.

Although complementation can be used for verification, it is not always practical. For complementation to work, it requires that either the phenotype is absent or destroyed (mutated) in a host. This is not always possible since genes coding for critical metabolic or regulatory enzymes can not be destroyed without killing the host. Similarly, it is nearly impossible to mutate a gene in a diploid organism (i.e., an organism with two of each chromosome). To successfully destroy a gene in a diploid cell would require two mutations, at the same exact location on each chromosome. The probability of this double mutational event is extremely low.

12.3   EXPERIMENTAL DESIGN AND PROCEDURES

Very often a putative (suspected) clone turns out to be false. It is not uncommon for oligonucleotides to hybridize to the wrong gene and yield a false positive. Verifying a clone's identity is therefore a mandatory step in the cloning process. Involved in the verification process is the elucidation of a restriction map, the determination of the gene's location on a fragment, the nucleotide sequencing of the gene, and, ideally, the introduction of that gene into a host where it is expressed and the product can be assayed. The following experiments are representative of the techniques used in the characterization and verification process.

Experimentally, up to this point you have been manipulating DNA in a manner which is not dependent upon the actual gene you are cloning. However, many of the remaining experiments are very dependent upon the DNA you have cloned. The probability that you have actually cloned *MEL*1 is low since the amount of work associated with that task far exceeds your responsibility. If you did clone the gene, feel proud. If you didn't, feel no shame. The *MEL*1 gene can be purchased through the American Type Culture Collection (ATCC 53360) and can be successfully used in place of your own clone.

Restriction Mapping

Restriction mapping employs several techniques covered earlier in this manual, namely restriction endonuclease cleavage of DNA and agarose gel electrophoresis. To demonstrate restriction mapping, you will digest your clone and lambda DNA (as a control) with several enzymes and electrophorese the products in order to determine the relative position of those restriction sites within the molecule. This same gel can also be used for the Southern blotting experiment.

Materials

Lambda DNA (1 μg/μl)

*Eco*RI

*Eco*RI 10X reaction buffer—500 mM Tris, pH 7.5, 100 mM MgCl₂, 1 M NaCl, 10 mM dithioerythritol

*Hind*III

*Bam*HI

*Hind*III and *Bam*HI 10X reaction buffer—100 mM Tris, pH 8, 50 mM MgCl₂, 1 M NaCl, 10 mM 2-mercaptoethanol

Lambda-*Hind*III marker (digoxigenin labelled)

*MEL*1 clone—isolated or purchased

Method

1. Set up the reactions as described in Table 12.1. Incubate the digests for 60 min at 37°C.

   Three of the above reactions require more than one restriction endonuclease. In this instance simultaneously using more than one enzyme to cleave the lambda DNA will work; however, it is important to note that multiple digestions are not always successful.

2. In three separate reactions, cleave 1 μg of plasmid (clone) with the three restriction endonucleases. Also prepare a negative control (i.e., no enzyme). The reactions should also be incubated for 60 min at 37°C. Each reaction should contain the following:

__.__	μl	ddH$_2$O
2.0	μl	10X reaction buffer
1.0	μl	restriction enzyme*
__.__	μl	plasmid DNA (1 μg)
20.0	μl	Total

Table 12.1 Components for Restriction Mapping of Lambda DNA

Component	*Hind*III	*Bam*HI	*Eco*RI	*Hind*III/ *Bam*HI	*Hind*III/ *Eco*RI	*Eco*RI/ *Bam*HI
Water	16 μl	16 μl	16 μl	15 μl	15 μl	15 μl
Buffer	2 μl	2 μl	2 μl	2 μl	2 μl	2 μl
Lambda	1 μl	1 μl	1 μl	1 μl	1 μl	1 μl
*Hind*III	1 μl			1 μl	1 μl	
*Bam*HI		1 μl		1 μl		1 μl
*Eco*RI			1 μl		1 μl	1 μl
TOTAL	20 μl	20 μl	20 μl	20 μl	20 μl	20 μl

*The restriction enzyme should be at a concentration of 1–2 U/μl. The negative control should lack enzyme.

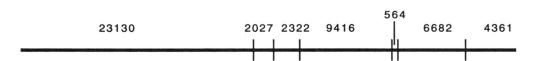

Figure 12.13 ■ Lambda-*Hind*III restriction map.

3. While the restriction digests are incubating, prepare a 1% agarose gel in 1X TAE or 1X TBE buffer. The volume of the agarose solution is dependent upon the type of electrophoresis unit being used.

4. Heat the lambda digests to 65°C for 10 min to separate the cos sites and then chill on ice. Add loading dye (1 µl of dye to 5 µl of digest) to the restriction digests, load onto the agarose gel (group the lambda DNAs and plasmid DNAs separately), and run the gel until the bromophenol blue tracking dye has traveled two thirds the distance of the gel or as long as time permits.

5. Stain the gel in an ethidium bromide bath for 10–30 min. Destain in water for 10 min. Observe the gel on a UV transilluminator and photograph. Save the gel for use in the Southern blot experiment.

6. The lambda DNA *Hind*III restriction map is shown below (Figure 12.13). Using this map and the photograph of the gel, determine the location of as many *Bam*HI and *Eco*RI restriction sites as possible. This is a puzzle (i.e., mind teaser) and must be pieced together with care and patience. The location of all the sites are not necessarily apparent.

7. From the plasmid's electrophoretic pattern, what information can you deduce about your clone?

Southern Blotting

Southern blotting is analogous to the colony blot (Chapter 11) except that DNA is digested with a restriction endonuclease and separated by agarose gel electrophoresis prior to its transfer to a blotting membrane.

By fragmenting the DNA, a target sequence can be identified since the probe will only hybridize to a homologous sequence located on a specific fragment. In this manner, Southern blotting is used to locate a target sequence within a larger DNA molecule.

Different techniques are used to transfer the DNA from the gel to the membrane, such as capillary, electro-, and vacuum blotting. In this exercise, traditional capillary blotting will be applied.

Materials

Agarose gel containing digested DNA—prepared previously

Nylon membrane

Probe—Digoxigenin-labelled oligonucleotide or random primed DNA prepared in Chapter 11.

Denaturation solution—0.5 N NaOH, 1.5 M NaCl

Neutralizing solution—1 M Tris, pH 7.5, 2 M NaCl

20X SSC transfer solution—0.3 M Na citrate, pH 7, 3 M NaCl

2X SSC—Diluted with water from 20X SSC

2X SSC, 0.1% SDS

0.2X SSC, 0.1% SDS

Wash Solution—100 mM Tris, pH 7.5, 150 mM NaCl

Glass baking dish

Napkins

Filter paper

Blocking solution—2% Blocking Powder* or BSA in 100 mM Tris, pH 7.5, 150 mM NaCl

Prehybridization solution—5X SSC, 1.0% Blocking Reagent,† 0.1% N-lauroylsarcosine, 0.02% sodium dodecyl sulfate. 20X SSC is 0.3 M Na citrate, pH 7, 3 M NaCl.

MORE...

---

*Item is supplied with Genius™ Kits

†Blocking Reagent is a 10% solution of Blocking Powder (Boehringer Mannheim Biochemicals) in 100 mM Tris, pH 7.5, 150 mM NaCl. First combine the Tris and NaCl, and then add the Blocking Powder. Warm the solution to dissolve the Blocking Powder. Do Not Boil. Excess Blocking Reagent can be stored frozen to prevent spoilage.

Heat sealable plastic bags (hybridization bags)

Heat sealer/impulse sealer

Substrate buffer—100 mM Tris, pH 9.5, 150 mM NaCl, 50 mM MgCl$_2$

Anti-digoxigenin antibody alkaline phosphatase conjugate*

Method

1. Soak the gel in denaturing solution (0.5 N NaOH, 1.5 M NaCl) in a small container for 15–30 minutes. The denaturing solution will diffuse through the gel and denature the DNA to a single-stranded form. Pour off the denaturing solution (this can be reused).

2. Neutralize the gel by soaking in neutralizing solution (1 M Tris, pH 7.5, 2 M NaCl) for 15–30 minutes. Pour off the solution (this can also be reused).

3. For capillary blotting, stack the following materials in the order listed in the glass baking dish:

    ▪ 0.5 cm of filter paper (cut slightly larger than the gel)

    ▪ agarose gel (bottom up, remove bubbles between paper and gel)

    ▪ nylon membrane (remove bubbles between membrane and gel)

    ▪ several pieces of filter paper (again remove bubbles)

    ▪ 4–5 inches of napkins

    ▪ a heavy, hard cover book (slightly larger than the napkins)

4. Add 20X SSC transfer solution to the baking dish up to the bottom of the gel. Capillary action will pull the transfer solution up through the gel and into the napkins. The DNA will move with the solution, but will be captured by the nylon membrane (which is positively charged). Transfer the DNA overnight at room temperature.

---

*Item is supplied with Genius™ Kits

5. Remove the membrane from the stack and wash it in 2X SSC to remove residual agarose. At this point, bake or UV link the DNA to the membrane. UV linking is accomplished by placing the membrane face down on the UV transilluminator and irradiating for 3 min. Baking is done for 2 hr at 80°C. If nitrocellulose membranes are used, baking must be done in a vacuum oven so that the membrane does not combust due to the presence of oxygen.

6. Place the membrane in a heat-sealable bag. Leaving the top of the bag open, seal around the membrane to minimize the volume of the bag. Leave 1–2 inches between the opening of the bag and the top of the membrane. Add prehybridization solution to the bag in a volume of 200 µl per cm² of membrane and heat seal along the top of the bag. Remove bubbles from the bag before sealing. The prehybridization solution contains protein that will bind to the charges on the surface of the membrane; this prevents the probe from nonspecifically sticking to the membrane. Prehybridize the membrane for 1–2 hrs at 68°C for random primed probes and T_d–5 for oligonucleotides.

7. If a random primed probe is to be used, denature the probe by heating it in a boiling water bath for 10 min. Place the probe on ice immediately and until use. Open the bag and add the probe directly to the prehybridization solution. Use 5–20 ng/ml of prehybridization solution for digoxigenin labelled probes. If an oligonucleotide probe is used, no denaturation is necessary since the probe is single-stranded. Simply add 1–10 pmoles of oligonucleotide/ml of prehybridization solution. When adding either probe, take care not to touch the membrane with the pipette tip (this will cause background).

8. Remove bubbles from the bag, seal, mix, and incubate overnight in a waterbath (use the same temperature as prehybridization). Oligonucleotide probes can actually be incubated for as little as 1–6 hours.

9. Open the bag and pour off the hybridization solution. This solution can be saved (frozen) and reused several times (up to five times for a DNA probe). Hybridization solutions that contain double-stranded probes must first be boiled to denature the probe before reuse.

10. Wash your membrane for 10 min with 2X SSC, 0.1% SDS in the bag at room temperature. When sealing the bag, it is important to remove

air bubbles and avoid membrane drying, both of which can create background. Repeat this wash.

11. Wash the membrane at 68°C (or T_d–5 for oligonucleotide probes) for 15 min in the waterbath with preheated 0.2X SSC, 0.1% SDS. Washing at elevated temperatures can be done by sealing the solution and membrane in a plastic bag and floating the bag in a water bath. Repeat this washing step.

12. Equilibrate the membrane in Wash Solution (100 mM Tris, pH 7.5, 150 mM NaCl) for 1 min.

13. Open the bag and add 200 $\mu l/cm^2$ of blocking solution (2% blocking powder, 100 mM Tris, pH 7.5, 150 mM NaCl) and incubate for 1–2 hrs at room temperature. Avoid exposing the membrane to air bubbles during the incubation.

14. Open the bag and add 1:5000 anti-dig alkaline phosphatase conjugate to the blocking solution (e.g., 1 μl per 5 ml). Do not touch the pipet tip against the membrane. Incubate for 1 hr at room temperature. Avoid exposing the membrane to air bubbles during the incubation.

15. Open the bag and pour off the antibody solution. Wash with Wash Solution twice for 15 min each wash. Avoid drying the membrane.

16. Equilibrate the membrane with alkaline phosphatase substrate buffer (100 mM Tris, pH 9.5, 150 mM NaCl, 50 mM $MgCl_2$) for 1 min. *Do not allow the membrane to dry out!* The membrane is now ready for either the colorimetric or chemiluminescent visualization.

Colorimetric Detection

Materials

Wash Solution—100 mM Tris, pH 7.5, 150 mM NaCl

Substrate buffer—100 mM Tris, pH 9.5, 150 mM NaCl, 50 mM $MgCl_2$

Substrate solution—X-Phos and NBT* in substrate buffer

*Item is supplied with Genius™ Detection Kits.

Method

1. Prepare substrate solution by adding 45 µl NBT (nitroblue tetrazo-lium at 100 mg/ml in DMF) and 35 µl X-Phos (5-bromo-4-chloro-3-indoyl-phosphate at 50 mg/ml in DMF) into 10 ml of substrate buffer (100 mM Tris, pH 9.5, 150 mM NaCl, 50 mM $MgCl_2$). This substrate is light sensitive so avoid prolonged exposure to the light. Unused substrate solution may be stored frozen.

2. Pour off the substrate buffer and add substrate solution (100 µl/cm^2). Seal the bag and incubate in the dark for up to 12 hours.

3. When the color is sufficiently developed, wash the membrane with Wash Solution or water to stop the reaction. The blot can then be dried and stored in a sealed bag. It is best to photocopy or photo-graph the blot shortly after visualization since the color typically fades with time (i.e., photobleaches).

Chemiluminescent Detection

Materials

> Substrate buffer—100 mM Tris, pH 9.5, 150 mM NaCl, 50 mM $MgCl_2$
> Lumi-Phos™530
> Kodak XAR Film
> Sterile transfer pipettes
> Acetate page protector
> X-ray cassette
> GBX developer
> Stop bath
> Kodak Rapid Fixer

Method

1. When visualizing a gel by chemiluminescence, it is extremely impor-tant not to let the membrane dry during its preparation. After soaking

the membrane in substrate buffer, remove the membrane from its bag and lay it between the pages of a clear acetate page protector (i.e., the type sold by stationery stores). Using a pipette, cover the membrane with several drops of substrate buffer to prevent it from drying out.

2. The substrate Lumi-Phos™530 will be used to detect the alkaline phosphatase tagged probe. The alkaline phosphatase catalyzes the removal of a phosphate from the Lumi-Phos™530 which creates an unstable intermediate. The intermediate emits light as it stabilizes. Lumi-Phos™530 can easily be contaminated by omnipresent alkaline phosphatase so take care not to contaminate the bottle of Lumi-Phos™530. We recommend that the substrate be aliquoted to smaller clean tubes (sterile, disposable). If clear tubes are used, wrap them with foil to protect the substrate from light. Using a sterile transfer pipette, cover the membrane with Lumi-Phos™530, e.g., 4–5 drops per 5 × 10 cm blot. Close the page protector.

3. Using several Kimwipes, spread the Lumi-Phos™530 by wiping the acetate. Drain excess Lumi-Phos™530 substrate from the acetate. Remove any moisture from the surface of the acetate.

4. In a darkroom, place the acetate on X-ray film, e.g., Kodak XAR, in the X-ray cassette and incubate for 90 min. Light emission will steadily increase for 6–8 hours. The following day an equivalent exposure will take approximately 10 min. The membrane can be reexposed over the next several days if different exposures are desired.

5. Develop the film. If developing trays or tanks are used, we suggest:

GBX Developer	2 min
Stop Bath	30 sec
Rapid Fixer	10 min
Water Rinse	15 min

6. Air dry the film.

Evaluating the Southern Blot

The film or membrane, depending on the visualization technique employed, should reveal a lambda-*Hind*III ladder and at least one band. By comparing the mobility of the positive band to the migration of the lambda fragments, it is possible to determine which fragment contains the clone. Compare the blot to the original photograph of the gel. Which band on the photograph hybridized to the probe?

Sequencing of the *MEL*1 Gene

One method of clone verification is to determine its nucleotide sequence. The predominant technique used for DNA sequencing is the Sanger dideoxynucleotide termination method. This method generates a population of terminated DNA molecules that represents every base in the DNA sequence. This population is then separated by polyacrylamide gel electrophoresis to yield a pattern representative of the nucleotide sequence.

DNA sequencing requires extensive preparation of biomolecules and reagents. The target DNA (i.e., template DNA) must be produced and purified, the oligonucleotide primers must be synthesized (or purchased) and labelled, the reagents must be prepared, the apparatus must be carefully washed and assembled, and then the gel must be mixed and poured. The actually dideoxy chain termination can be performed once the template is purified.

There are many variations available for dideoxy chain termination sequencing. Without reinventing the past, we recommend a technique known as cycle sequencing. Cycle sequencing evolved from the polymerase chain reaction (covered in Chapter 13) and involves recycling the template DNA for multiple sequencing reactions. Basically a template, termination mixture (i.e., dNTPs and one ddNTP), Taq polymerase, and labelled primer are (1) heated to 95°C to denature the template; (2) cooled to a suitable primer annealing temperature; and (3) increased to 70°C for polymerization (the optimal temperature of activity for Taq polymerase). This process is then repeated 20 or more times. This termination reaction is then separated on an acrylamide gel.

Materials

Dig Taq DNA Sequencing Kit (Boehringer Mannheim)

5' digoxigenin labelled sequencing primer (included in kit)—The two primers provided in the kit are specific for pUC plasmids and derivatives. These primers are on either side of the multiple cloning site and sequence into the MCS.

Sequencing apparatus

Recombinant plasmid with insert—cleaned

Phenol:chloroform:isoamyl alcohol (25:24:1)—Under TE buffer.

3 M potassium acetate, pH 5.5

95% ethanol, ice cold

TE buffer—10 mM Tris, pH 8, 1 mM EDTA

Nylon membrane

Blocking solution—2% Blocking Powder in 100 mM Tris, pH 7.5, 150 mM NaCl

Wash Solution—100 mM Tris, pH 7.5, 150 mM NaCl

Lumi-Phos™530

Page protector acetate sheets

GBX developer

Stop bath

Rapid fixer

X-ray film

Film cassettes

Thermocycler

Method—Template Preparation

1. The plasmid DNA needs to be free of contaminating protein. If the plasmid is not protein free, extract the DNA with an equal volume of phenol:chloroform:isoamyl alcohol. Take precautions when handling phenol and chloroform. Mix the aqueous and organic phases and separate by centrifugation. The aqueous layer is on the top.

2. Transfer the DNA solution to a new microfuge tube and add $\frac{1}{10}$ volume of potassium acetate. Mix and add two volumes of ethanol. Precipitate the DNA at –20°C for 30 min or longer.

3. Centrifuge the DNA in a microfuge at maximum speed for 15 min. Decant and dry the DNA pellet. Clean DNA may be very difficult to see as a pellet.

4. Resuspend the DNA in TE buffer to a concentration of 1 µg/µl. The volume will be determined by the initial mass of DNA. For the sequencing reaction, an aliquot of DNA should be diluted to 10 ng/µl in TE buffer.

Method—Cycle Sequencing

1. Prepare four tubes with 4 µl of the A, T, G, and C termination mixtures provided with the cycle sequencing kit.

2. In a separate tube, prepare the following reaction:

Primer (Dig-labelled—2 pmoles)	2	µl
10X cycling mix (with Taq polymerase)	4	µl
Template DNA (10 to 100 fmoles)	—	µl
Water	—	µl
Total	20	µl

3. Dispense 4 µL of the solution made in step 2 into each of the four termination solutions prepared in step 1. Cap each reaction with a small volume of mineral oil, i.e., 10–20 µl. The oil is thick, and exact measurement is difficult. The idea is to seal the reaction to prevent evaporation. If the thermocycler has a heated cap or bonnet, then the oil is not necessary.

4. Cycle the reaction 20 times under the following conditions:

    95°C for 1 min
    T_d–5°C* for 1 min
    70°C for 1 min

---

*This T_d is specific for the oligonucleotide primer. For the pUC sequencing primer T_d–5 is 47°C while for the reverse sequencing primer T_d–5 is 45°C.

5. Following the thermocycling, run the reactions on the sequencing gel. Since the termination products are nonradioactive, they may be stored at –20°C until time permits for their analysis.

Method—Apparatus Preparation

1. The assembly of a sequencing apparatus is specific for the type of unit you have available. Therefore, follow the manufacturer's instructions about its assembly. There are certain general guidelines, however, which should be followed so that the gel is prepared correctly.

2. The glass plates of a sequencing gel must be thoroughly cleaned prior to the assembly of the apparatus. Wash the plates with soap and water (or ammonia cleaner) and rinse well. Lay the plates on paper towels clean side up. Wash the plates with 95% ethanol and Kimwipes. Allow the plates to dry and then repeat the washing.

3. It is generally a good idea to silanize one of the two plates so that when the plates are separated later, the gel will only adhere to the unsilanized plate. In a well ventilated area, pour a small amount (3–4 ml) of Sigmacote (from Sigma) onto one of the glass plates and spread it over the plates with a Kimwipe. Allow the plate to dry. The plates are now ready for assembly.

Gel Preparation and Pouring

Casting a sequencing gel is much the same as pouring a small gel for protein analysis (Chapter 6), the major differences being the gel mixture, the type of comb, and the size of the gel itself. The method presented here is simple but illustrative of the process.

Materials

Acrylamide solution (40%)—19:1 acrylamide/bis-acrylamide
Urea
5X TBE buffer
TEMED

MORE...

Ammonium persulfate—freshly prepared 10% solution

25 ml disposable pipettes

Sharktooth comb

Assembled sequencing plates

Method

1. One 20×40 cm sequencing gel requires less than 50 ml of acrylamide solution; however, a stock solution is usually made and stored at 4°C until needed. First dissolve 230 g of urea in 200 ml of water (warming the solution may help in the dissolution). Add 75 ml of the 40% acrylamide solution and 50 ml of 5X TBE. **Caution: Acrylamide is a neurotoxin!** Add water so that the total volume is 500 ml. The final concentration of the acrylamide is 6%. This can be stored refrigerated for several months.

2. Place 50 ml of 6% acrylamide solution in a 250 ml sidearm flask. Degas the acrylamide by covering the top with a rubber stopper and applying a vacuum (some researchers consider degassing optional). Once the solution stops bubbling (i.e., the gas stops coming out of solution), add 200 µl of 10% ammonium persulfate and 50 µl of TEMED. Mix the solution by swirling. Pour the gel immediately since the polymerization can occur rapidly.

3. Tilt the gel apparatus up at a 30° angle from the lab bench. Using a 25 ml pipette, add the acrylamide solution to one corner of the plates. Tilt the plates as needed so that the solution fills the plates evenly without trapping bubbles. Fill until the solution is 0.5 cm from the top of the shorter plate.

4. Insert the shark tooth comb backwards into the acrylamide. The even edge of the comb will form an even layer of polymerized gel. Lower the gel so that comb is only a couple of centimeters above the lab bench. Plastic wrap can be placed over the comb and top of the gel to prevent drying. Allow the gel to stand for 1 hr or until polymerization has occurred (keep your eye on the extra mix in the flask as an indicator of polymerization).

5. Once polymerized, assemble the complete gel apparatus. Remove the shark tooth comb, invert, and place it into the gel. The teeth of

the comb should just enter the gel. Once inserted, do not move the comb. Add 0.5X TBE to both the upper and lower buffer reservoirs. The gel is ready to load and run.

Electrophoresing and Visualizing the Sequencing Reactions

Loading a sequencing gel is very similar to loading a protein gel. The samples are mixed with a formamide buffer and heated to separate the termination products from their templates. The sample is then pipetted into the wells. The wells are unusual in that they are defined by the teeth of the sharktooth comb. Once loaded, a voltage is applied to separate the termination products.

Materials

Assembled sequencing unit

Dideoxynucleotide termination solutions—prepared earlier

Formamide loading buffer—90% formamide, 20 mM EDTA, 0.3% bromophenol blue, 0.3% xylene cyanol.

Micropipettes with gel loading tips

Heat block

Method

1. To each of the termination solutions, add 2 µl of formamide loading buffer. Mix and heat to 95°C for 5 min. Chill on ice.

2. Using gel loading tips, flush any bubbles and buffer from the areas between the teeth of the comb. Being consistent and precise, load all the contents of each termination reaction onto the gel. Load the samples in the sequence T C G A in adjacent wells.

3. Connect the leads and run the gel at 35 watts until the lower dye reaches the bottom of the gel.

4. Visualizing the sequencing gel is very similar to the protocol used for the Southern blot. The first step is to disassemble the sequencing apparatus, but *do not separate the glass plates.*

5. Place the glass plates onto paper towels covering the lab bench. The plate which was *not* silanized should be on the bottom. Remove any clamps and tape holding the plates together.

6. Cut a nylon membrane so it is slightly larger than lanes which are to be visualized. Do not touch the middle of the membrane with your hands. Only handle it from the edge or corners.

7. Remove the top glass plate by prying the plates apart with the tip of a flat spatula. The gel should stay on the lower plate. If the gel sticks to the upper plate, simply switch the plates.

8. Hold the membrane at the top and bottom edges but with your hands nearly touching. The membrane should be hanging in a loop below your hands. Touch the membrane to the center of the gel and layer it outward and evenly over the gel. *You will only have one chance to lay this membrane on the gel correctly.* Place a piece of filter paper on top of the membrane. Using a pipette like a rolling pin, gently remove any bubbles between the membrane and the gel. Place the top glass plate on the filter paper. Place a thick book or catalog on top of the plate. Allow the termination products to transfer to the membrane for 30–60 min.

9. Remove the book, plate, and filter paper. Carefully peel the membrane off the gel. Any residual acrylamide that sticks to the membrane can be removed by washing with Wash Solution (100 mM Tris, pH 7.5, 150 mM NaCl).

10. Fix the DNA to the membrane by cross-linking. Place the membrane face down on the UV transilluminator and irradiate for 3 min. The membrane may be cross-linked in stages if it is larger than the UV screen.

11. Place the membrane in a heat sealable bag and equilibrate the membrane in Wash Solution (100 mM Tris, pH 7.5, 150 mM NaCl) for 1 min. Open the bag and pour off this solution.

12. Add 200 μl/cm$_2$ of blocking solution (2% blocking powder, 100 mM Tris, pH 7.5, 150 mM NaCl) and incubate for 1–2 hr at room temperature. Avoid trapping air bubbles in the bag.

13. Open the bag and add 1:5000 anti-dig alkaline phosphatase conjugate to the blocking solution (e.g., 1μl per 5 ml). Do not touch the pipet tip against the membrane. Incubate for 1 hr at room temperature. Again, avoid trapping air bubbles in the bag.

14. Open the bag and pour off the antibody solution. Wash with Wash Solution twice for 15 min each wash. Avoid drying the membrane.

15. Equilibrate the membrane with alkaline phosphatase substrate buffer (100 mM Tris, pH 9.5, 150 mM NaCl, 50 mM MgCl$_2$) for 1 min. The membrane is now ready for either the colorimetric or chemiluminescent visualization. The visualization protocols for the Southern blot (Chapter 12.3) are also used for this experiment. Visualize the DNA by following the Chemiluminescent Detection protocol.

16. Save the film for analysis in Chapter 13.

Transformation of Yeast with Cloned DNA

The most definitive means of verifying a clone's identity is to introduce that clone into a new host and to find its product (i.e., its protein). To verify a clone in this manner, the host must lack the gene or its resulting protein. Often a host, such as *S. cerevisiae*, will be mutated to destroy the gene which is to be subsequently introduced. If the gene is naturally absent, then introducing the cloned gene is less complicated.

A gene introduced into a new host will not function properly unless its regulatory sequences are recognized. If the promoter is not recognized, than a host promoter must be substituted for the clone's promoter. Though this heterologous gene may not be expressing naturally, the resulting protein should be native, i.e., functional. This allows for verification of the proper end product. However, recombinant proteins produced in foreign hosts do not always function correctly. The host may alter the protein or possess a cellular environment that prevents proper protein folding or enzyme activity. This obstacle is very difficult to overcome; however, if the protein is to be detected by an immunological assay, protein activity is not necessary.

In this experiment, you will introduce the *MEL1* gene into a *mel⁰* strain of *S. cerevisiae*. Since the donor and new host are compatible (actually *S. carlsbergensis* is classified as a subspecies of *S. cerevisiae* by many taxonomists) *MEL1* should function well in its new environment. In reality, this experiment represents a large jump forward in the research process. For instance, the *MEL1* gene would first need to be subcloned into a yeast shuttle vector, i.e., a vector that replicates in both *E. coli* and *S. cerevisiae*. For instructional purposes, we suggest using the specific ATCC clone or equivalent.

Materials

Yeast culture*

Yeast plasmid with *MEL1*—Cloned DNA or ATCC 53360, i.e., plasmid pMEL.D.9.1. This plasmid contains the *MEL1* gene on a yeast shuttle vector that also has the *URA3* gene. Thus, this plasmid can be introduced and selected in yeast that contain a *ura3* mutation.

YPD medium—Sterile; 25 ml in a 125 ml flask plugged with cotton

TE buffer, 10 mM Tris, pH 8, 1 mM EDTA

Li acetate buffer—100 mM Li acetate, 10 mM Tris, pH 8, 1 mM EDTA

50% polyethylene glycol

Sorbitol buffer—1 M sorbitol, 10 mM Tris, pH 8, 1 mM EDTA

Uracil minus, selective agar plates—Per 100 ml, 0.67 g Yeast Nitrogen Base, 1 g galactose , 2 g agar, 120 mg adenine sulfate, 360 mg leucine

Method

1. Inoculate a flask of YPD broth (2% glucose, 2% peptone, 1% yeast extract) with an isolated yeast colony from a freshly streaked YPD plate. Incubate the culture with agitation, overnight at 30°C.

---

*Yeast strains which can be used for this experiment must be carefully selected. The strain must not only be *mel⁰*, but also be capable of metabolizing galactose. Common laboratory strains of yeast, such as DBY 746 and DBY 747, possess *MEL1* though it is not indicated in genetic descriptions. Strains that lack *MEL1* often contain mutations in the galactose metabolic genes. We commonly employ strain BTC 1958, which possesses a mutated *MEL1* gene (i.e., *mel1*). This strain is available through the Biotechnology Training Institute.

2. Pellet yeast from 4.5 ml of broth by centrifuging for 5 min at 3000 rpm.

3. Decant and then resuspend the yeast in TE buffer and centrifuge again at 3000 rpm for 5 min.

4. Decant and then carefully resuspend the yeast in 3 ml of 100 mM Li acetate buffer and shake gently at room temperature for 30 min. As calcium causes *E. coli* to become competent, Li acetate does the same for the yeast.

5. Centrifuge the yeast for 5 min at 2000 rpm. Carefully remove the supernatant and resuspend the yeast in 100 μl of Li acetate buffer and transfer to a 1.5 ml microfuge tube. The density of the yeast is approximately 5×10^8 cells/ml.

6. Add up to 10 μg of plasmid DNA to the yeast in a volume not greater than 10 μl. Incubate at 30°C without shaking for 30 min. The amount of DNA that is used to transform yeast is significantly higher than that used for *E. coli*. Where 1 μg of plasmid can yield in excess of 10^5 *E. coli* transformants, 10 μg of plasmid may only transform 100–200 yeast.

7. Add 300 μl of 50% polyethylene glycol (PEG), mix and incubate at 30°C for 1 hr without shaking. The PEG is extremely viscous and significant amounts may stick to pipette tips.

8. Heat shock the cells for 5 min at 42°C. Like *E. coli*, this heat shock is a critical factor for a successful transformation.

9. Immediately centrifuge the cells for 20 sec. Remove the PEG solution and resuspend the cells in 1 ml of sorbitol buffer.

10. Spread plate the cells on selective media. The selective agar plates contain Yeast Nitrogen Base, supplemented with leucine, and galactose. The amino acid that corresponds to the complementing gene on the plasmid is omitted. Cells that develop on this agar should contain the plasmid due to their growth on the minimal medium. If *MEL1* is present, the galactose should induce the synthesis of α-galactosidase. Allow the plates to dry before inverting and incubating.

11. Colonies will take several days to grow. These colonies should be picked and assayed for α-galactosidase activity. This is done by mixing 25 µl of p-nitrophenyl-α-D-galactosidase, 25 µl of water, and 50 µl of 0.5 M acetate buffer, pH 4.5. Inoculate this reaction with a small portion of a yeast colony and incubate for 5 min. Quench the reaction with 3 ml of 0.1 M sodium carbonate. A yellow color indicates the presence of the *MEL1* gene and verifies the identity of the clone.

STUDY QUESTIONS

1. Using a plasmid such as pUC119 (Figure 12.1), suggest a method to specifically clone a fragment from a larger clone. You can assume the larger clone has been restriction mapped (e.g., Figure 12.6).

2. Investigate the meaning of RFLP. How does RFLP relate to the topics in this chapter?

3. What are two practical applications of the Southern blot?

4. Investigate the differences between Sanger dideoxy chain termination sequencing and the Maxam and Gilbert chemical degradation sequencing.

5. Devise a strategy for sequencing DNA without previously cloning the DNA.

FURTHER READINGS

Henikoff S (1984): Unidirectional digestion with exonuclease III creates targeted breakpoints for DNA sequencing. *Gene* 28:351–359

Innis M, Myambo K, Gelfand D, Brow M (1988): DNA sequencing with *Thermus aquaticus* and direct sequencing of polymerase chain reaction-amplified DNA. *Proc Nat Acad Sci USA* 85:9436–9440

Maxam A, Gilbert W (1977): A new method for sequencing DNA. *Proc Nat Acad Sci USA* 74:560–564

Sambrook J, Fritsch EF, Maniatis T (1989): *Molecular Cloning: A Laboratory Manual*. Plainview, NY: Cold Spring Harbor Laboratory Press

Sanger F, Nicklen S, Coulson A (1977): DNA sequencing with chain terminating inhibitors. *Proc Nat Acad Sci USA* 74:5463–5467

Southern E (1975): Detection of specific sequences among DNA fragments separated by gel electrophoresis. *J Mol Biol* 98:503–517

Chapter

13

Application of DNA Sequence and Clone Data

13.1 OVERVIEW

The sequence of nucleotides within a DNA molecule can yield a wealth of information. Important biological features such as amino acid sequence, exact amino acid composition, and gene structure are contained within sequences. The data can be used for highly specific manipulations of cloned DNA, such as changing a single nucleotide, substituting promoters between genes, and fusing of genes to yield hybrid (heterologous) proteins. Probes can also be designed for a variety of uses, including the detection of genetic diseases and pathogenic microbes.

The nucleotide sequence of a clone is fundamental. From this data, all known restriction sites, subsequences, and coding regions can be determined. The comparison of sequences allows for the elucidation of important biological motifs, such as the Pribnow box, the alteration of sequences through site-directed mutagenesis, and gene fusion methodologies. The synthesis of homologous oligonucleotides allows for the design of highly specific probes (or primers). These oligonucleotides, or primers, have led to the development of an extremely powerful technique, the polymerase chain reaction (PCR). Based on the specificity of primers, PCR allows for the in vitro replication of small regions of DNA.

The objective of the preceding twelve chapters was to gather data from a protein and clone its associated gene. Whether you were successful in this endeavor or simply followed the process, that objective has been achieved. In reality, the means by which you cloned the *MEL1* gene, however, is not important. It is the gene itself, including its nucleotide sequence, that is important.

Now that you have a gene and its sequence, what do you do with it? This chapter will focus on the analysis of DNA sequence data and on one method for its application.

13.2 BACKGROUND

The cloning and characterization of DNA are prerequisites for its study and application. Although the emphasis of the preceding chapters dealt with how to clone, it is the clone characteristics and sequence data itself that are important. By comparing and contrasting DNA sequences, biologically important subsequences can be defined. A subsequence is usually a short stretch of bases that performs an important biological function. These subsequences are not absolutely uniform but contain bases that have a high rate of occurrence within the location.

Powerful molecular analysis software is available that allows for sequence evaluation, and includes such major programs as GCG (Genetics Computer Group). Alternatively, more convenient software programs are available for the personal computer (e.g., MacVector™). Some of the information elucidated from nucleotide sequences is listed in Table 13.1.

Probe Design Using Molecular Analysis Software

Undoubtedly computers have become an extremely powerful tool for the molecular biologist. Although the comparison of DNA sequences generates useful information, molecular analysis software can be used at many stages during the discovery process. Perhaps the first use of analysis software could have occurred as far back as Chapter 7 for the design of an oligonucleotide probe based on an amino acid sequence of

Table 13.1   Information Obtainable from DNA Sequence Data

Item	Sequence	Comment
Open Reading Frame	ATG* NNN$_n$ STOP	STOP = TAA, TAG, TGA
Shine-Dalgarno Sequence	AGGAGG(N)$_{6-9}$DTG	D is any base but C.
Pribnow Box	TATAAT	Prokaryotic −10 promoter consensus sequence.
−35 Region	TTGACA	Prokaryotic −35 promoter consensus sequence.
Polyadenylation Sequence	AATAAA	Eukaryotic mRNA tailing signal.
Exon/Intron Splice Site	MAG/GTPuAGT	Pu is either G or A. M is A or C. The / is the location of the splice.
Intron/Exon Splice Site	(Pu)$_{\geq 11}$NPyAG/G	Py is a C or T. N is any base. The / is the location of the splice.
CAAT Site	GGPyCAATC	Eukaryotic promoter element. Py is C or T.
Condon Usage and Bias	Protein Specific	Analysis of coding sequence which shows number and frequency of codons used for protein coding.

*Occasionally, GTG and TTG can act as the Start condons. Eukaryotes do not always start with ATG.

a protein. Software, such as MacVector™, can efficiently reverse translate an amino acid sequence and then provide suggestions about the sequence coding for the least degenerative oligonucleotide probes. For instance, the N-terminal sequence for the α-galactosidase is:

> Met Phe Ala Phe Tyr Phe Leu Thr Ala Cys Ile Ser Leu Lys Gly
> Val Phe Gly Val Ser Pro Ser Tyr Asn Gly Leu Gly Leu Thr Pro
> Gln Met Gly Trp Asp Asn Trp Asn Thr Phe Ala Cys Asp Val
> Ser Glu Gln Leu Leu Leu

Reversed translated it is:

> ATG TTY GCN TTY TAY TTY YTN ACN GCN TGY ATH
> WSN YTN AAR GGN GTN TTY GGN GTN WSN CCN WSN
> TAY AAY GGN YTN GGN YTN ACN CCN CAR ATG GGN
> TGG·GAY AAY TGG AAY ACN TTY GCN TGY GAY GTN
> WSN GAR CAR YTN YTN YTN

Key: A = adenine, C = cytosine, G = guanine, T = thymine, N = A,C,G, or T, H = any base but G, W = A or T, S = G or C, and R = A or G.

Using MacVector™, analysis for potential DNA probes 17 to 18 base pairs long yields the following options.

There is one probe 17 base pairs long containing eight permutations (i.e., variation based on degeneracy). Y is either T or C.

AA #	34	Trp	Asp	Asn	Trp	Asn	Thr
Base #	100	TGG	GAY	AAY	TGG	AAY	AC

	min	max
G + C content:	35.3%	52.9%
Tm:	46.0°C	52.0°C

Additionally, there are three probes 18 base pairs long, with 16 permutations each. These probes are all extracted from the same region.

AA #	32	Met	Gly	Trp	Asp	Asn	Trp
Base #	94	ATG	GGN	TGG	GAY	AAY	TGG

	min	max
G + C content:	44.4%	61.1%
Tm:	52.0°C	58.0°C

AA #	32	Met	Gly	Trp	Asp	Asn	Trp	Asn
Base #	95	TG	GGN	TGG	GAY	AAY	TGG	A

	min	max
G + C content:	44.4%	61.1%
Tm:	52.0°C	58.0°C

AA #	32	Met	Gly	Trp	Asp	Asn	Trp	Asn
Base #	96	G	GGN	TGG	GAY	AAY	TGG	AA

	min	max
G + C content:	44.4%	61.1%
Tm:	52.0°C	58.0°C

These oligonucleotides would be applicable as probes for genomic cloning. The degeneracy of the probe can be overcome by synthesizing mixed sequences on a DNA synthesizer. For instance, the 17 base oligonucleotide after synthesis would actually be a pool of eight different probes.

Basic Sequence Analysis

Another use of molecular analysis software involves the construction of a full length DNA sequence from individual sequencing reactions. A Sanger dideoxy DNA sequencing reaction will generate several hundred termination products; however, due to the limitations of electrophoresis, less than 300 bases (i.e., termination products) can usually be read on a gel. This limit in the amount of readable sequence obtainable requires the adoption of strategies that decipher the entire sequence in bits and pieces. Several strategies are used for elucidating long sequences, such as shotgun and walking approaches.

Shotgun sequencing involves digesting a clone into many small pieces and then ligating those pieces into a vector. The result is a population of subclones, all of which in total represent the original clone. Using a primer that is specific for the vector, and anneals adjacent to the inserted subclone, the population of subclones can be sequenced. The resulting data are generated in a random fashion and as such do not apparently overlap. By using sequence aligning software (e.g., AssemblyLIGN™), these sequences can be arranged according to their overlapping regions (Figure 13.1). Shotgun sequencing was once very common, but today has been surpassed by the more elegant walking strategy.

Sequencing by walking is a directed approach, and it is performed in discrete steps that start by sequencing from the vector into an insert. Once the insert DNA sequence is analyzed, a new primer is synthesized to the 3' end of that sequence. This new primer is used to continue the sequencing from that location. As new sequence is continually generated, new primers are designed and synthesized to continue the analysis (Figure 13.2). In this manner, each sequence serves as a stepping stone for the continued sequencing of adjacent regions.

Once a sequence has been determined and assembled, it is useful to identify the more basic characteristics. Two immediate computer searches

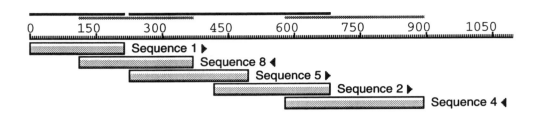

Figure 13.1 Alignment and direction of fragments produced by shotgun sequencing by computer analysis with AssemblyLIGN™.

can locate restriction endonuclease recognition sites and open reading frames (ORFs). These are usually simple analyses since the variation in these sites is minimal. For instance, a restriction endonuclease usually cleaves a very specific site, such as GAATTC by *Eco*RI. The uncovering of a restriction site on a sequence is definitive since the enzyme has a specific recognition site. Sequence analysis software typically has a database of restriction enzymes which can be updated and modified. The software searches this database and then displays a restriction enzyme map (Figure 13.3).

With prokaryotic DNA, open reading frames (ORFs) are also simple to detect. Most ORFs start with the codon for methionine, namely ATG. Since amino acids are determined by three nucleotides, i.e., the codon, the examination of each codon past an ATG until a stop codon (i.e., TAA, TAG, or TGA) is reached constitutes an open reading frame (Figure 13.4).

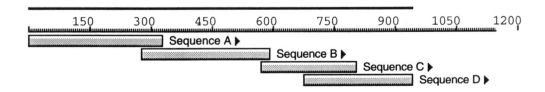

Figure 13.2 DNA sequence overlap and alignment produced by walking. Analysis and alignment made use of AssemblyLIGN™ software.

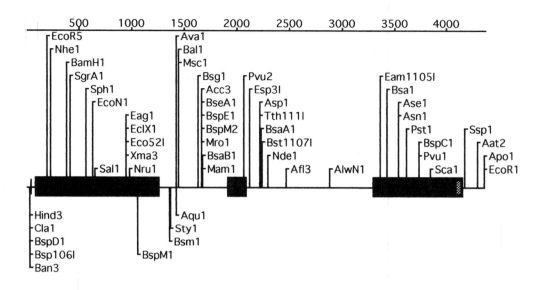

Figure 13.3 A MacVector™ generated map of unique restriction endonuclease sites on pBR322 as deduced from the DNA sequence. The black bars represent plasmid encoded genes.

(These open reading frames are only possible protein coding regions.) The identification of open reading frames in eukaryotes is not easily determined since coding regions may include intervening sequences, or introns. These introns disrupt the triplet codon pattern between the start and stop codons.

If the molecular weight of a protein is known, then translated open reading frames can be compared to the protein's mass. As such, the amino acids that compose the ORF can be used to verify the clone based on the protein data. Rapid, specific determinations of molecular weight can be deduced and compared to experimental data, such as protein size as measured by SDS-PAGE. The one drawback in this comparison is that molecular analysis software cannot judge the mass associated with glycosyl groups attached to many extracellular proteins. The correct ORF can also be determined by matching the N-terminal protein sequence with codons on the DNA sequence.

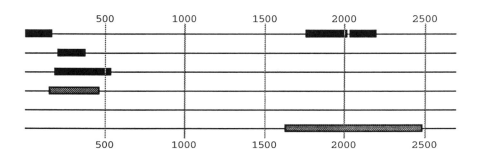

Figure 13.4   An open reading frame map from the plasmid pUC19.

The comparison of DNA sequences has generated and continues to generate a tremendous amount of information about genes and how they function. For instance, the analysis of numerous *E. coli* promoters revealed two relatively conserved sequences centered around 10 and 35 bases in front of the location where transcription is initiated. The examination of over 100 *E. coli* promoters provides an estimate of the frequency of specific bases in the promoter and allows for the creation of a consensus sequence (Figure 13.5). This extraction of biologically important sequences, also known as subsequences, is a continuing process and a powerful tool built into analysis software.

Sequences can also be compared to reveal homologous regions. Depending on the extent of homology, a matrix analysis assesses the relatedness of two DNA sequences. In the Pustell matrix analysis, the sequences are compared, and homologous regions are plotted as dots or lines on a X–Y graph (Figure 13.6). This is a visual way to observe homologous regions. Lines of homology that appear on a Pustell matrix may indicate a biologically important or conserved region, such as an enzyme active site.

A sequence can also be compared to enormous databases of known sequences, such as GenBank and EMBL. This can be extended to the analysis of a specific probe designed to hybridize only to its homologous sequence. Such probes have many applications, including the

	-35 Region	-10 Region
Consensus	TTGACA	TATAAT
araC	TGGACT	GACACT
trp	TTGACA	TTAACT
recA	TTGATA	TATAAT
Lambda P_L	TTGACA	GATACT
pBR322 RNA I	TTGAAG	TACACT

Figure 13.5 The *E. coli* consensus sequence and other promoter sequences.

Scoring Matrix: DNA identity matrix

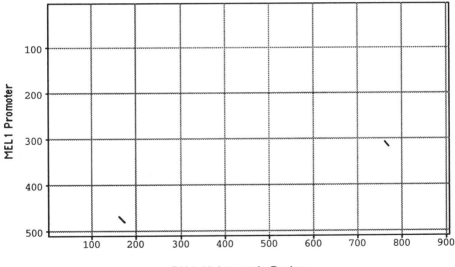

Figure 13.6 Sequence homology as determined by a Pustell matrix analysis. The two short lines indicate homologous sequences between *MEL*1 and *GAL*1–10 promoter regions. This homology is significant since both promoters are part of the galactose regulon. Both promoters interact analogously with a common DNA-binding, regulatory protein. Analysis was performed with MacVector™.

detection of pathogens and determination of genetic diseases. The specificity of the probe minimizes the probability of nonspecific hybridization.

Application of Sequence Data

Sequence data can be used for a multitude of applications, such as site-directed mutagenesis, a precise technique used to alter one or more specific nucleotides within a sequence. By designing a probe with a mismatched center, but homologous flanks, an imperfect hybridization can be performed between the probe and its target (Figure 13.7). Heat denaturing the DNA followed by annealing of the misaligned primer provides sufficient hybridization to allow for the synthesis of the homologous strand with DNA polymerase (e.g., Klenow fragment). In this manner, the DNA can be replicated in vitro while including the minor alteration. The alternative is random mutagenesis using physical and chemical agents, such as UV light and nitrosoguanidine.

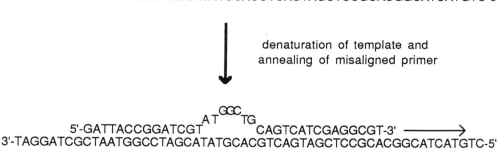

Figure 13.7   Site-directed mutagenesis using misaligned oligonucleotide primers.

A major technique that applies DNA sequence data is the polymerase chain reaction (PCR), a process in which DNA is amplified exponentially in vitro. PCR makes use of the instability of the double helix at high temperatures, the specific hybridization of oligonucleotides, and the action of DNA polymerases.

The polymerase chain reaction can be divided into three steps: (1) denaturation; (2) annealing; and (3) extension. The denaturation of DNA is usually accomplished by heating the DNA to 95°C for a period of one minute or less. This temperature is normally sufficient to denature most DNA molecules. Annealing occurs at a temperature that is slightly below the dissociation temperature (T_d) of the oligonucleotide. An oligonucleotide has a T_d that is dependent upon the length and composition of the molecule. As the length and/or G/C content of a primer increases, so does its T_d. This increase is obviously not infinite since full length DNA can be denatured at 95°C. A typical annealing temperature for a primer is between 37° and 70°C. The extension of the primer is accomplished using a thermostable DNA polymerase, such as Taq DNA polymerase isolated from *Thermus aquaticus*, a resident of hot springs. This enzyme can tolerate the heat generated during the repeated denaturations. The extension occurs at 70–72°C, the optimum temperature for Taq polymerase, and usually for a duration of one minute. All three steps together comprise a cycle, and a typical PCR will have between 25 and 40 cycles. This cycling causes an exponential increase of DNA. If one of the two primers is limiting, as in cycle sequencing (Chapter 12.3), then the increase in DNA occurs at a linear rate.

The key to PCR lies in the fact that two oligonucleotides (primers) are designed to anneal to opposite strands of DNA so that their 3' hydroxyl groups point towards each other (Figure 13.8). The extension (polymerization) of the primers replicates the homologous binding sequence for the other primer. The new double helices are then denatured, new primers are annealed, and then again extended thus recreating the binding sequences. By the end of the cycling, a few original DNA molecules can be amplified to yield enormous quantities of molecules.

PCR is used extensively in molecular biology for DNA detection, site-directed mutagenesis, and measurement of gene expression. PCR can be applied to the detection of pathogens, specific alleles (i.e., recessive genetic traits), and more. Its application to gene expression analysis first requires the synthesis of cDNA (via reverse transcriptase) followed by amplification, this overall technique being termed rtPCR ("rt" for reverse transcriptase). This technique has also been adapted to amplify DNA and RNA within tissues and cells, i.e., in situ PCR and in situ rtPCR. Such measurements can show the genetic make-up of individual cells, such as lymphocytes carrying HIV. PCR is both rapid and sensitive, with a typical reaction taking less than three hours and being capable of amplifying 10,000 molecules or less to μg amounts of DNA.

PCR relies on primers that are specific to the DNA to be amplified. As such, PCR first requires a fair amount of up front research, specifically the elucidation of the sequence to be amplified. To use PCR for the clinical identification of antibiotic resistant bacteria, for example, would first require the cloning and sequencing of the antibiotic resistance genes from the pathogen. Once the DNA sequence is deciphered, specific primers for PCR can be designed.

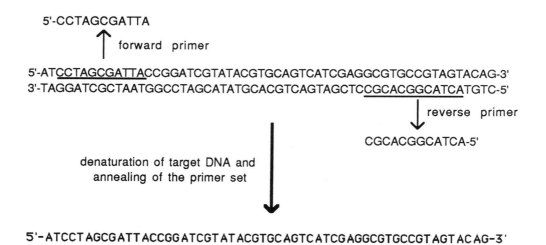

Figure 13.8　Alignment of primers on DNA for PCR.

13.3 EXPERIMENTAL DESIGN AND PROCEDURES

DNA sequence data in itself is not valuable until it is analyzed and, if possible, also applied. The laboratory portion of this chapter will focus on two aspects: (1) the analysis of a DNA sequence for biological information; and (2) the amplification of DNA via PCR.

Computer Analysis of DNA Sequence

A simple computer analysis of a nucleic acid sequence can unearth extensive information and detail. The location of restriction endonuclease recognition sites alone is extremely valuable if the DNA is to be subcloned. The identification and translation of open reading frames can be used to locate the coding region of a gene while the sequence itself can be used as a means of confirming the composition and amino acid sequence of a protein. More complex analyses can reveal protein structural motifs and DNA regulatory domains.

Analysis of DNA sequence does not require the purchase of expensive computer software. Most colleges and universities have access to molecular analysis software, such as GCG (Genetics Computer Group), through a mainframe or shareware programs. An inexpensive shareware program, such as DNA Strider, can provide valuable information about a sequence (especially for demonstration purposes). MacVector™ is an excellent analysis program available for the Macintosh personal computer. This software, which is featured in this manual, is capable of extensive DNA and protein sequence analysis. The objective of this session is to use a computer to analyze a DNA sequence and extract information.

Materials

Computer with DNA sequence analysis software

Method

1. If a DNA sequence is not available, the following sequence can be entered. This sequence is of the *MEL1* gene as determined by Sumner-Smith et al. (1985).

```
GAATTCTTTCTGTACGCTCAGGGTGGGCCTTTAAAGGATAGC
ACCCTACCGAAGTCGACTTCTAAGTAAACACCATTACTAGGA
GATGACTAAATCTGGAAAACACATGGTGGTCTGAATGCGTCT
AGTCTCTGCCATAAACATAACATGTTTGTTTTAATGCATTCT
CGTGTTTAATCGACATTAATGTGGGGGGGAGAAAGACATCCCA
TCCCTGAAAGGTTTTTCCAGGGAATAGTCAGGACGCATTGGC
TTTCATTCGGCCATATGTCTTCCGAAAGAAGAAGAAAGGAAG
ACATGTATTACATTATCCAACAAAAAATGGTTCTTGACGTCT
ACAAATCAAGAATCTTAAAGACATTGAACGAAGTAGCTGAA
TAAAAATTATGAAAACTATAAAAACTATAAAAACTGTACTT
AAGTCCTCAATAAAACATAAACTTCTTACTGTATAAGGTTT
TCGATAATTTCTTACTTGATTCTAGGAGAGCAACGGTAATAA
AAGCAACGACGATGTTTGCTTTCTACTTTCTCACCGCATGCA
TCAGTTTGAAGGGCGTTTTTGGGGTGTCTCCGAGTTACAATGG
CCTTGGTCTCACTCCACAGATGGGTTGGGACAACTGGAATAC
GTTTGCCTGCGATGTCAGTGAACAGCTACTTCTAGACACCGC
TGATAGAATTTCTGACTTGGGGCTAAAGGATATGGGTTACAA
GTATATCATTCTGGATGACTGCTGGTCTAGCGGCAGAGATTC
CGACGGTTTCCTCGTTGCAGATGAACAAAAATTTCCCAATGG
TATGGGCCATGTTGCAGACCACCTGCATAATAACAGCTTTCT
TTTCGGTATGTATTCGTCTGCTGGTGAGTACACCTGTGCTGGA
TATCCTGGGTCTCTGGGTCGTGAGGAAGAAGATGCACAGTTC
TTTGCAAATAACCGCGTTGACTACTTGAAGTACGATAATTGT
TACAATAAGGGTCAGTTTGGTACACCGGAAATTTCTTACCAC
CGTTACAAGGCCATGTCAGATGCTTTGAATAAAACTGGTAGG
CCTATATTCTATTCTCTATGTAACTGGGGTCAGGATTTAACA
TTTTACTGGGGCTCTGGTATCGCCAATTCTTGGAGAATGAGT
GGAGATGTTACTGCTGAGTTCACTCGTCCAGATAGCAGATGT
CCCTGTGATGGCGATGAATACGATTGCAAGTACGCCGGTTTC
CATTGTTCTATTATGAATATTCTTAACAAGGCAGCTCCAATG
GGGCAAAATGCAGGTGTTGGTGGTTGGAATGATCTGGACAAT
CTAGAGGTTGGTGTCGGGAATTTGACTGACGATGAGGAAAAG
GCACATTTCTCTATGTGGGCAATGGTAAAGTCTCCACTTATC
ATTGGTGCCAATGTGAATAACTTAAAGGCATCTTCGTACTCA
ATCTATAGTCAAGCCTCTGTCATCGCAATTAATCAAGATTCA
AATGGTATTCCAGCAACAAGAGTCTGGAGATATTATGTTTCA
GACACAGATGAATATGGACAAGGTGAAATTCAAATGTGGAG
```

MORE…

```
TGGTCCTCCTGACAATGGTGATCAAGTGGTTGCTTTATTGAAT
GGAGGAAGCGTATCTAGACCAATGAACACGACCTTGGAAGAG
ATTTTTTTTGACAGCAATCTGGGTTCAAAGAAACTGACATCG
ACTTGGGATATCTACGACCTATGGGCCACCAGAGTTGACAAC
TCGACAGCGTCTGCTATCCTTGGACGGAATAAGACAGCCACC
GGTATTCTCTACAATGCTACGGAGCAATCCTACAAAGACGGT
TTGTCTAAGAATGATACAAGACTGTTTGGTCAGAAAATTGGT
AGTCTTTCTCCAAATGCTATACTTAACACGACTGTTCCAGCT
CACGGTATCGCCTTCTATAGGTTGAGACCCTCTTCTTGAGCTT
ATTGTTGAGCAAAGCAGGGCGAGAAGTATTGATGATTGTTAA
AAAGTTCATGAAAAAAATACTACTCGAATATTTATTCAGAG
TAACTAAATAATAAACGACAGAATAGCCTATCAGGTATTCC
AATAGTTTTCGTTTTGTAGGTACATAATCTGAAGCCCTTGAA
CTTTTTCTCGTTTACATACTTCATTGCATTAGCGATATTTCA
CATGTGCTATAC
```

2. Analyze the sequence for restriction endonuclease recognition sites and open reading frames. The steps for accomplishing this search will be specific for the software you have available. For details, consult the appropriate reference guide.

3. Other features that can be examined should include:

   a. hydrophilicity—regions of the protein that are either hydrophobic or hydrophilic

   b. protein subsequences—biologically important protein domains

   c. DNA subsequences—biologically important DNA domains

   d. translated ORF—amino acid code for putative protein

4. Within this sequence, can you identify an ORF (open reading frame) that represents α-galactosidase?

Amplification of DNA by the Polymerase Chain Reaction

The polymerase chain reaction is an extremely powerful tool for the detection and manipulation of DNA. PCR allows for the amplification of specific sequences as designated by the primer pair. This amplification has been used for the detection of numerous DNAs including those associated with pathogens and genetic diseases. In this exercise, you will simply amplify *MEL1* DNA directly from yeast genomic DNA.

Materials

Template DNAs—10^5 to 10^6 copies/10 μl in TE buffer

Primers—50 pmoles/μl each (forward and reverse)

10X reaction buffer—500 mM KCl, 100 mM Tris-HCl (pH 8.0), 20 mM MgCl$_2$

dNTP's (dATP, dCTP, dGTP, dTTP) 10 mM stocks of each or combined

Taq DNA polymerase—5 U/μl

Sterile mineral oil

Sterile pipette tips

Sterile distilled H$_2$O

Programmable heat block or thermocycler

Method

1. You will be amplifying the yeast *MEL1* structural gene. Based on the nucleotide sequence provided above, 13 primer pairs were derived through the MacVector™ sequence analysis software (Figure 13.9). As defined by the computer search, each set of primers generates a PCR product which is between 300 and 600 base pairs. The individual primers have also been specified to be between 18 and 22 bases in length, have a T$_m$ of 55–80°C, and contain 45–55% G+C. Two of these primer pairs are below (Figure 13.10). For this PCR, either primer set can be used.

2. Add the following reagents (in the order listed) to a *0.5 ml microfuge tube*. Adding of dNTP's to unbuffered water may result in the hydrolysis of phosphate esters, thus it is important to mix the water and buffer first. Furthermore, PCR is susceptible to amplifying contaminating DNA molecules, consequently, use a different pipette tip for the addition of each reagent. Never enter a reagent stock solution with a used pipet tip.

Tm (°C): 55 - 80
percent G+C content: 45 - 55
product size: 300 - 600
3' dinucleotide: NS

primer vs. primer (G-C only): 2
3'-end vs. 3'-end: 2
3'-end vs. product: 5

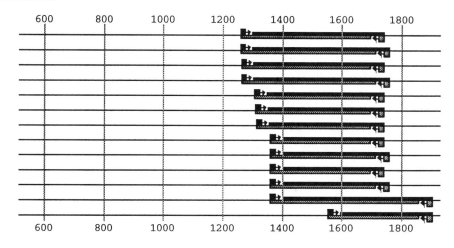

Figure 13.9   Potential PCR primer pairs for *MEL*1 as determined by MacVector™.

The reaction will contain the following:

__.__ μl	distilled H₂O*	

___.___ μl	distilled H_2O*
10.0 μl	10X reaction buffer
8.0 μl	dNTP's (2 μl of each)
___.___ μl	Forward primer (50 pmoles)*
___.___ μl	Backward primer (50 pmoles)*
10.0 μl	template DNA
0.5 μl	Taq DNA polymerase
100.0 μl	Total

*Adjust the volume of the water and primers so that the entire reaction is 100 μl.

Forward Primer 1: 1259-1278 5'–GGGGCAAAATGCAGGTGTTG–3'
   20 nucleotide forward primer, pct G + C: 55.0, T_m: 57.5
Backward Primer 1: 1742-1721 5'–AAGGATAGCAGACGCTGTCGAG–3'
   22 nucleotide backward primer, % G + C: 54.5, T_m: 55.3
   484 nucleotide product for this pair (1259–1742)
   Optimal annealing temp: 54.9, % G + C: 42.8, T_m: 76.0

Forward Primer 2: 1262-1282 5'–GCAAAATGCAGGTGTTGGTGG–3'
   21 nucleotide forward primer, % G + C: 52.4, T_m: 57.6
Backward Primer 2: 1761-1740 5'–TGGCTGTCTTATTCCGTCCAAG–3'
   22 nucleotide backward primer, % G + C: 50.0, T_m: 55.7
   500 nucleotide product for this pair (1262–1761)
   Optimal annealing temp: 55.1, % G + C: 42.8, T_m: 76.1

Figure 13.10 PCR primer sets and characteristics.

3. Overlay the reaction with 75 µl of sterile mineral oil. The oil is extremely viscous and requires care in pipetting. The oil will prevent evaporation of the reaction mix during the heating steps. If a thermocycler with a heated lid is used, then the mineral oil is unnecessary.

4. Place a small drop of mineral oil in each well of the thermocycler to be used (this will allow the heat to be conducted evenly from the block to the tube). Place the 0.5 ml tubes into the thermocycler. Program the thermocycler with the following cycle parameters: denaturation at 95°C for 1 min, annealing-specific for primer set for 1 min, and extension at 70°C for 1 min. Repeat for 35 cycles. Start the reaction.

5. Following the reaction, store the tubes at 4°C or analyze immediately.

6. Several assays may be used to detect the amplified product. We will examine your PCR products by agarose gel electrophoresis after the amplification to assess the results. This electrophoresis should be

performed with 2% agarose and as previously described (see Chapter 9). The gel should reveal a single sharp band.

STUDY QUESTIONS

1. Two yeast genes that code for histidine biosynthetic enzymes show a stretch of homology in their promoter regions (as determined by a Pustell matrix). What experiments could you devise in order to test whether these homologous regions are important for the regulation of their associated genes?

2. Develop a strategy and assay approach using PCR for the detection of genetic diseases, such as cystic fibrosis. Consider including a step in your assay which verifies the result.

3. Investigate codon bias and determine what effect it may have on expressing a gene in a foreign host.

FURTHER READINGS

Erlich H, ed. (1989): *PCR Technology*. New York: Stockton Press

Innis M, Gelfand D, Sninsky J, White T, eds. (1990): PCR Protocols. San Diego: Academic Press

Reznikoff W, Gold L, eds. (1986): Maximizing Gene Expression. Boston: Butterworth

REFERENCES

Sumner-Smith M, Bozzato M, Skipper N, Davies R, Hopper J (1985): Analysis of the inducible *MEL1* gene of *Saccharomyces carlsbergensis* and its secreted product, α-galactosidase (melibiase). *Gene* 36:333–340

von Heijne G (1987): Sequence Analysis in Molecular Biology—Treasure Trove or Trivial Pursuit. San Diego: Academic Press

ADG-6287 10/22/96

Sci Lib
QP
551
B95
1995

RESERVE BOOK

DATE ON:

CH 313 ~~SPRING RESERVE~~ ~~1998~~

Burden & Whitney
BIOTECHNOLOGY:
    PROTEINS TO PCR